Readers Praise *MindBody Ca*

"In the spring of 1999, I had the pleasure of working with Dr. Morry Edwards and using his book. I have personally seen the lives of many patients improved by listening to and reading Dr. Edwards' book. It is a model of old-fashioned common sense and the latest in mind-body research. Throw into that mix topics like biofeedback, journaling and the spiritual life, and the reader has just moved from the fear and isolation of cancer to the holistic and comforting world of healing. Dr. Edwards really is a pioneer in the integration of medicine into the mind-body world. I hope that this book is welcomed and read with joy by professionals and patients alike."

—Carl Bell
Psycho-Oncology Counselor
Goshen General Hospital
and Cancer Survivor

"Over his many years of counseling cancer patients, Dr. Morry Edwards has developed perceptive insights into how different people react to living with cancer. Based on these collective personal experiences, he has given us a set of useful suggestions for proactively living with cancer. Many breast cancer survivors have already benefited greatly from learning and practicing the principles and concepts presented by Dr. Edwards. We are especially indebted to him for his perseverance and dedication to helping cancer patients not only survive but thrive."

—Rosemary (Bunny) LaDuke, President
Southwest Michigan Affiliate
Susan G. Komen Breast Cancer Foundation
and Cancer Survivor

"This book provides the patient diagnosed with cancer a much needed resource tool to enhance health and healing. Morry Edwards presents well-researched information in a practical, easy to use format, covering a variety of topics, from coping, to embracing life. His presentation of psychoneuroimunology translates into strategies and techniques that are easy to implement for the reader to use at home, or with a leader in a group setting. He also provides the reader with numerous resources for further reading. I highly recommend this book as a valuable tool for patients, their family and friends, and practitioners interested in health, wellness, and healing."

—Marcia Prenguber, N.D.
Director of Integrative Care, Center for Cancer
Care, Goshen Health System

"Every cancer patient and their family should have this manual. Our cancer program has relied on Dr. Edwards' lessons for several years to help patients manage the stress of a cancer diagnosis and treatment. It has made an impact; our patient satisfaction scores rank within the top 5 percent in the country."

—R. Paulette Brown, CMPE
Executive Director
Center for Cancer Care,
Goshen Health System and
—Richard B. Hostetter, MD, FACS, Medical
Director and Surgical Oncologist

"Morry, I think the *MindBody Cancer Wellness* book is great; I know I'll refer to it often. I was especially grateful to be in the class both times. Have a great new year!!"

—Marlene Koch, Cancer Survivor

"...out of the despair, a bright light. Morry Edwards, Ph.D. I'd met him years before. As an attorney, after receiving excellent recommendations about his scholarship, his honesty and abilities, I asked him to be an expert witness... When I found I had cancer, thank God, Morry was there in the cancer support group lighting the way—a way to deal with cancer. To have hope, to push the darkness away. Many of the exercises in the workbook were used in our group. They teach you to look at yourself and others, to grab each day. To have hope...The manual brings me back out of the abyss. When I start slipping, the manual comes out. I re-read and re-work the exercises. They work. This is the basic 'get your stuff together' book... After attending several support groups, I can say I've seen so many people just devastated by cancer. They are shattered. This book brings facts and ideas together to help manage the unmanageable—the unthinkable. I recommend it heartily to people to get 'back on track.'...I've lived with cancer for more than three years. Thanks to Morry and *MindBody Cancer Wellness*, I am living and loving and filling every moment with as much life as I can. I am putting a positive light into my life. Thank you, God....Thank you, Morry."

—Lynda Lowry, Cancer Survivor

"...I was introduced to Morry Edwards' *MindBody Cancer Wellness workbook* in a stress management class for breast cancer patients. This book presents a wide variety of very valuable techniques for meditation and mental exercises to aid in maintaining a positive attitude during the stressful months of dealing with cancer. I find I also utilize these techniques in my present wellness. All of us encounter stressful times in our everyday lives; this books assists in an optimistic and holistic approach to the inevitable ups and downs of life, especially during cancer treatment."

—Barbara Z. Cooley, Cancer Survivor

"I'd like to thank you for all the things you have taught me since my diagnosis of breast cancer in 1997. Being aware of the physical indicators, my body tells me when to slow down and relax to keep the tension level from stress from spiraling upward. My life since the diagnosis is so much better than before. I laugh more, enjoy people more, and am more aware of the gifts of friendship and love. The subsequent diagnosis of lung cancer this last year and how I was able to handle it, I directly credit to your teachings. Cancer does not define who I am. Thank you so much for your care and what you have taught me. Thanks for being my friend."

—Patricia (Pat) MacLeod, Cancer Survivor

This is a book that EVERYONE, cancer patient or not, needs to read. Morry Edwards is a pioneer in mind-body techniques and biofeedback, with twenty-plus years of compassionately counseling cancer patients. This manual weaves his special insight with empowering and transformative techniques to give us the most healing ingredient of all: HOPE.

—Lynda Kirk, MA, LPC,
BCIA Senior Fellow, QEEGT
Clinical Director, Austin Biofeedback Center

MindBody

Cancer Wellness

A Self-Help Stress Management Manual

Morry D. Edwards, Ph.D.

MindBody Cancer Wellness: A Self-Help Stress Management Manual

Cover Photo: © Barbara Armstrong
Limited Text Graphics: PrintMaster Gold®
Editing: Dr. Diane Hamilton, Professor, WMU School of Nursing
   and Dawn M. Edwards, Research Associate; Administrative Assistant, Wellness Psychology®
Printed in the United States of America
First Edition, 2000. Second Edition, 2003.

   Library of Congress Cataloging-in-Publication Data

Edwards, Morry.
  Mindbody cancer wellness : a self-help stress management manual /
Morry D. Edwards.-- 2nd ed., rev. and enlarged.
     p. cm.
  ISBN 0-9728969-0-2
  1. Cancer--Alternative treatment. 2. Stress management. 3.
Cancer--Palliative treatment. 4. Psychoneuroimmunology. 5. Holistic
medicine. I. Title: Mind body cancer wellness. II. Title.

  RC271.A62E385 2003
  616.99'406--dc21
                                    2003005124

Acorn Publishing
A Division of Development Initiatives
P.O. Box 84
Battle Creek, MI 49016-0084
http://www.acornpublishing.com

ISBN 0-9728969-0-2

DEDICATION

This manual is dedicated to the memory of my parents. My father, Abe Edwards, died forty years ago after a starburst grew in his brain. He was given three months and lived eighteen, which in those days was a miracle. He taught me that a life's path could develop out of the most tragic of circumstances. This book is also dedicated to my mother, Edith Wolfson Edwards, who died after an eight-year battle with breast cancer. She taught me that an act of will could prolong life until our purpose is accomplished. This book is also dedicated to the memory of the thousands of people I have counseled over the last twenty-five years who have been touched with cancer. There have been tremendous lessons and coming into people's lives at this intense time always humbles me. It is also dedicated to a future of hope to ease the journey of many who will have to travel their path with cancer.

I would also like to acknowledge my beautiful wife who has been an unending source of support throughout this project. Her aesthetic sense helped develop the layout. Also, I want to acknowledge my son, Micah, who at seven years old has kept me young. I hope he will carry on and help unravel cancer's mystery so his generation will not suffer from this set of diseases.

TABLE OF CONTENTS

PREFACE

This self-care manual is intended to give you skills and strategies to ease the stress of cancer and its treatment. Many of the techniques are "generic" and can be applied by anyone to any stressful situation. This book is not intended to replace common sense, professional medical, nor psychological help. It is intended to help you become a more active participant in your treatment and reduce your anxiety, hopelessness, anger, and fear. This manual can be a major help to you and your medical team if there are mental or emotional stress concerns. Should you have severe, unusual, unexplained, or persistent symptoms, I encourage you to seek appropriate professional evaluation and treatment. Also keep in mind that any symptom need not necessarily reach an extreme before seeking help.

Should a symptom or problem persist beyond two weeks despite your use of these self-care recommendations, consult a professional. The material in this book should not conflict with professional advice that you receive. Your health care professionals likely know your specific history and individual needs. Above all, you know what is best for you. The techniques and suggestions offered in this manual should help you feel more relaxed, confident, comfortable, or energized. If your experience using these strategies is uncomfortable, then they may not be for you. You may need to consult a professional for additional guidance.

This manual supplies many excellent tools to strengthen your emotional and mental well-being. It is my attempt to introduce a wide variety of ideas and strategies. You do not need to feel you have to master every technique in this book. Hopefully, you will find at least three ideas that benefit you with continued practice. This book is only a guide. Your common sense, good judgement and partnership with your healthcare team are essential.

The information contained within this book is as accurate and up-to-date as possible. All strategies and suggestions are based on the latest literature and research. Now in its second edition, we hope to continue to revise and improve this book. Your feedback and suggestions are welcome. Please send them to WELLNESS PSYCHOLOGY, P.O. Box 402, Plainwell, MI 49080 or email them to phdpurple@aol.com. Thank you. May this manual help you on your healing path.

Morry Edwards, Ph.D.

USING THIS MANUAL FOR CANCER WELLNESS

Erma Bombeck once said that there was a time on this planet when cancer and optimism weren't compatible. As we cross the threshold to the 21st century, we are coming closer to cancer's molecular secrets, which will unlock the door to new treatments that will be less toxic.

The deliberate use of cancer and wellness together in the title would appear incompatible. Cancer has always had an intensely negative image and been that "Dread Disease." All the negativity of a death sentence and conjured pictures of wasting away with this terrible disease immediately swarm down on the newly diagnosed person. Particularly, if he or she has had a painful personal experience with a family member or friend, then one is likely to have a negative attitude, regardless of how many positive statistics are quoted. Likewise, there are people who have had a positive example and begun with a better attitude. Regardless of outlook, many of the thousands of people I have counseled over the years thought they would never be able to handle getting cancer, but there they were coping and discovering great strength and deep truths about themselves. The research also consistently shows that people who take an active role in their treatment and feel they are partners with their healthcare team tend to do better. They have fewer complications from treatment, they feel more in control, they comply better with their treatment, and they have an easier time adjusting. These strategies may even help them live longer.

People still have horrible experiences with cancer, but more and more, people (8 to 10 million, in fact) are not just surviving cancer, but thriving after their experience. As I began writing the first edition of *MindBody Cancer Wellness*, Lance Armstrong, a man who had testicular cancer that had spread to his lungs and brain, had just won the Tour de France, an extremely grueling bicycle race. After his victory, he said, "**I hope this sends out a fantastic message to all survivors around the world. We can return to what we were before—and even better**." Since that time, Lance has won four straight races. Other people with sudden reversals of their cancer are raising questions about what healing forces are tapped by attitude and beliefs, prayer, lifestyle change, and "will to live." Caryle Hirshberg and Marc Barasch call for a *Remarkable Recovery Registry* to assess patterns that may reveal those healing forces rather than passing them off as statistical anomalies or misdiagnoses.

Regardless of the physical outcome, cancer represents a crisis situation. With any crisis, there is an opportunity for growth and development as well as death and decay. Cancer is so traumatic that it can shake things up enough for people to evaluate their lives and make changes so that healing occurs in a number of areas. It can improve relationships and it can bring in to focus what is important.

The most common feeling I hear expressed is "I'm out of control." The diagnosis of cancer is still an overwhelming and traumatic experience. People with cancer need to feel empowered. This manual is a way for people to actively participate in their care and restore a sense of control. As stated earlier, research indicates that people adjust better and feel less stress if they feel in control and participate actively in treatment decisions. This manual presents ideas and strategies to do just that. You need not master everything; in fact, some of the techniques may not appeal to you at all. Try what you can and practice a technique for a designated time -- aim for one to two weeks. If the technique benefits you by creating calmness, increasing energy level, or producing a sense of peace, then continue using it. If it does little for you, try something else. The key is to find the right path for you. Usually, when someone is diagnosed with cancer everybody has something to try. There is no way to follow up on everything, so do things that make sense to you and from which you derive benefit.

People make a distinction between a physical cure and positive healing that extends to all aspects of a person's life. There is another distinction that is important between health, which is the absence of disease, and wellness, which extends to a positive and joyful state of being and prevention. It looks like the continuum below.

Disease   Health   Wellness

On the next page are 12 points to keep in mind to remain well during the course of cancer and beyond. May this book help you find the right treatment path and ease your journey in life. As Joseph Campbell once said, **"If there was already a path, it would have to be someone else's; the whole point is to find your own way."**

SOME IMPORTANT ELEMENTS OF A CANCER WELLNESS DOCTRINE
Morry Edwards, Ph.D.

Healing and health promotion of the whole person is vitally necessary despite the course a physical disease may take. Some important beliefs that enhance wellness during a cancer diagnosis are listed below.

1. There are a large number of health factors that I can control or modify in taking responsibility for my health. I will explore all areas that can improve my health.

2. Cancer, like any crisis, represents an opportunity to change my life in a positive direction. I didn't need it and I don't have to be a good sport about it! Having some intense feelings does not mean I am not coping.

3. While I cannot change my diagnosis, I can control my attitude and reaction, particularly in the way I handle stress and my emotions. The past is unimportant unless I make it so. The future has not happened yet. The only reality is being fully alive in the present.

4. A wide range of feelings and reactions to cancer is normal but I need to redirect my energy from unproductive emotions such as worry, anger, fear, and resentment, into acceptance, love, and healing.

5. A positive attitude refers more to dealing honestly with my feelings rather than maintaining a happy face. I will try not to avoid any of the feelings I experience.

6. I do not have to be a professional cancer patient. That is not my occupation and there are many other aspects to my life. I need to set constructive and realistic goals in all areas of my life.

7. I have the power to make a difference in my care. I need to look within myself for proper direction.

8. My doctor and I are partners. We both have things to learn. I will be comfortable and confident with the treatment path I choose.

9. Cancer is a social disease. Improving the quality of relationships can become a source of healing support for me. This also includes my spiritual relationships.

10. Acceptance is not giving in. As Norman Cousins said, "Don't deny the diagnosis. Try to defy the verdict." Keep in mind, statistics apply to groups, not to a given individual. At least one person has defeated each type of cancer.

11. There is always hope, but what I hope for may change over time. As Dr. Carl Simonton said, "In the face of uncertainty, there is nothing wrong with hope."

12. My personal dignity and quality of life are always the best measures of success. It takes only a split second to die and all the rest of my time goes to living as best and joyfully as possible. Death is not a failure.

Adopted and enlarged from David F. Cella, Ph.D., "Health Promotion in Oncology: A Cancer Wellness Doctrine," *Journal of Psychosocial Oncology*, Vol. 8(1) 1990, pp. 17-31.

Introduction

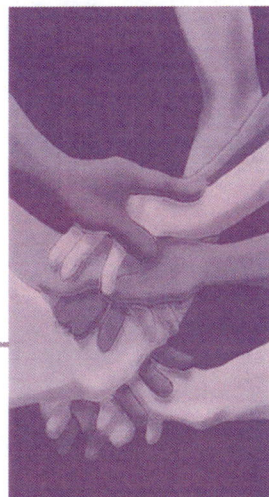

ALL I REALLY NEEDED TO KNOW ABOUT WELLNESS, I LEARNED FROM CANCER PATIENTS

MORRY EDWARDS, PH.D.
(With apologies to Robert Fulgham)

In a cartoon by Shanahan that appeared in *The New Yorker* in 1996, a doctor is relating good news to a snowman that his lumps were "just coal." Of course, we would like to hear these results from tests all the time. Unfortunately, three of the most dreaded words one can hear are: "You Have Cancer." You can't dress it up or put a "smiley face" on it. In 2003 approximately 1,334,100 people will hear those words. In Florida, where I was raised, 40,100 people will be diagnosed. In Michigan, where I live and work, 19,800 will be given this diagnosis. In Indiana, where I consult, 13,000 will learn they are facing a cancer diagnosis this year. And in Virginia, where my in-laws live, 13,700 will find out they have cancer.

On a more positive note, there are between 8 to 10 million cancer survivors. Cancer Wellness may seem like an oxymoron but throughout my career, working with this population has taught me as much about living and thriving as about dying. No one in his or her right mind would volunteer to contract cancer, but many people use it as a transformative experience, a chance to re-evaluate their priorities and make positive changes in their lives. I do not hear that only from the "lucky ones" who become long-term survivors, but many people undergoing treatment, who are often touched by the outpouring of love and acknowledgement they never knew existed.

It led one survivor to comment that having cancer changed her desires about the way she wanted to die. Sandy, diagnosed with advanced breast cancer, overheard two others talking at the funeral of a friend, "She was lucky to die so suddenly and quickly, not like having cancer or something that drags on." It immediately struck Sandy how wrong they were. She had wanted to die "with the quick strike" of a heart attack. Since her cancer diagnosis, she found that being directly confronted with her own mortality made her more appreciative of "the little pleasures" and brought much healing to family relationships. She was able to enjoy life much more until the end.

Some of the most valuable lessons I have learned by listening to my patients. They prompted me to develop at least 12 IMPORTANT PRINCIPLES OF CANCER WELLNESS, which actually could be applied to any chronic or life-threatening disease.

The following is an overview of these principles with grateful references to the real patients who brought them to life and have helped me understand their importance.

1. **There are a number of health factors that I can control or modify in taking responsibility for my health. I will explore all areas that can improve my health.**

    Monica was a thirty-year-old lymphoma patient who happened to be a psychotherapist friend of mine. She used multiple avenues to fight her cancer for a number of years. Besides doing guided imagery to improve her immune system, she made a number of lifestyle changes, including altering her diet, taking a regimen of vitamin supplements, and working through a number of personal and family issues. The most unique and powerful part of her wellness program was that she sculptured her tumor in clay. She started each day by shaving a small piece of her tumor off her "statue" which was a way for her to establish a positive attitude from the moment she awoke.

2. **Cancer, like any crisis, represents an opportunity to change my life in a positive direction. I didn't need it and I don't have to be a good sport about it! Having some intense feelings does not mean I am not coping.**

Lynda was a breast cancer patient in her early-fifties. She had been suffering and laboring and had been feeling drained for years as a lawyer. She was diagnosed with advanced disease after an extended workup. She decided to quit law and pursue her passion, which was art. That was the song she had wanted to sing for many years. Combining that major change and an integrated treatment program of conventional and alternative methods has kept her enjoying life for several years now.

3. **While I cannot change my diagnosis, I can control my attitude and reaction, particularly in the way I handle stress and my emotions. The past is unimportant unless I make it so. The future has not happened yet. The only reality is being fully alive in the present.**

Linda, a woman in her early-thirties, was diagnosed with breast cancer and discovered she was pregnant at the same time. She taught me early in my career how powerful mindbody techniques could truly be. Usually, I do not work with people until they manifest an obvious or serious problem and are referred from the oncologist or nursing staff. In this instance, the staff referred her preventatively because they were going to withhold chemotherapy for the baby's first trimester. Also, they were not going to give her any anti-emetic medication (medication which prevents vomiting) so as not to jeopardize the baby. They were already concerned that the anti-cancer drugs might harm the fetus. We worked together for six sessions. We did skin temperature and electromyographic biofeedback and relaxation-visualization exercises along with desensitizing her to the chemotherapy clinic by showing her what would happen with her treatments. I coordinated with the nursing staff to allow her about ten minutes with her meditation tape before they began her chemotherapy injections. She was a wonderful student and got so proficient at going into a relaxed state that she did not have any adverse post-treatment nausea or vomiting. The medical staff was astounded and it even surprised me a little bit. Now, Linda has a healthy teenager and has moved from the area.

4. **A wide range of feelings and reactions to cancer is normal but I need to redirect my energy from unproductive emotions such as worry, anger, fear, and resentment, into acceptance, love, and healing.**

Harriet, a melanoma patient in her late-fifties, taught me how destructive holding onto fearful and angry feelings could be. She was a perfect example of the "Damocles Syndrome" — "waiting for the shoe to drop" as the patients say — because she was always dreading the return of her cancer. Her cancer started as a small growth on her ear and spread to her brain. She had surgery and radiation and was in remission, meaning her cancer was non-detectable. She remained in this state for quite some time, but instead of reconnecting with her life, such as going places and healing relationships, she instead withdrew and fretted about a recurrence. Her cancer never returned, but unfortunately, she had a stroke that reduced her capacities considerably until she died. It always seemed that her death by dread, and not by cancer, came too early.

5. **A positive attitude refers more to dealing honestly with my feelings rather than maintaining a happy face. I will try not to avoid any of the feelings I experience.**

Ruby, in her late-forties, suffered from lymphoma. She taught me that some cancer patients feel a great deal of pressure to remain positive and become virtually prisoners of a positive attitude. No matter how terrible she may have felt, she would always say, "I'm fine," and try to smile. She would not even let down in therapy. There is a great deal of physical (not just emotional) stress in holding your feelings in like that. To keep ourselves under such tight control takes a great deal of energy, which diverts that energy from healing and wastes it. Ruby was trapped in her need to be positive, which may have contributed to her poor outcome.

6. **I do not have to be a professional cancer patient. That is not my occupation and there are many other aspects to my life. I need to set constructive and realistic goals in all areas of my life.**

Carolyn was a Hodgkins' lymphoma patient in her late-twenties. She taught me that some people have more to gain from their disease than by getting well. She got a lot of secondary gain from her disease in the way of attention and free service that she would never have received without her disease. Because she was emotionally needy, her twelve-year-old daughter became her parent. Instead of being a single mother, she became a victim who decided not to take conventional treatment so she valiantly fought the cancer establishment as well as the disease. She evoked a great deal of sympathy especially from her healthcare team whom she depleted one by one. She was afraid that her emotional needs would not have been met just being herself.

7. I have the power to make a difference in my care. I need to look within myself for proper direction.

Bill was a late-thirties colon cancer patient. He taught me that we all have a great deal of intuitive wisdom about what we need and that listening to our inner wisdom is vital to our healing. We discussed how important combining the rational information from outside professionals with our inner wisdom helps us to make the healthiest decisions. Bill was initially opposed to any chemotherapy because his mother had had such a difficult time with chemo for her breast cancer. Although I promised to support him either way he went, I asked him not to avoid chemo with fear as the only reason. He agreed and the next week returned with a reversed decision. While he went through chemotherapy, he also began making some positive changes in his life by getting back into running, changing his diet, improving relationships and becoming more assertive. I didn't hear from Bill for quite some time and after six months he came back to the office and had decided that the chemo had done some good but he now felt any further treatment would not be beneficial. I asked him to discuss this with his doctor, which he did. He and his doctor negotiated another month and then he stopped and that was more than twelve years ago. Since that time, research from clinical trials has confirmed that six months of chemotherapy is as effective as twelve months. Bill's intuition was right and it was healthy for him to pursue it. Our bodies, which house the subconscious, are very wise; we have to learn to quiet down long enough to hear that wisdom and follow it.

8. My doctor and I are partners. We both have things to learn. I will be comfortable and confident with the treatment path I choose.

Millie was a mid-seventies breast cancer patient who taught me how important direct and accurate information is in the prevention of needless suffering. She was a grandmotherly lady who initially came to the clinic with a very pleasant demeanor. The nurses playfully argued over who would treat her. She would bring in treats for the staff and always seemed to have something positive to say to the staff. Gradually, after four or five treatments, her whole attitude changed. She became increasingly irritated and would only let certain nurses touch her. She got so upset that she began to vomit in the parking lot before she even had her treatments. I was then asked to see her. When we sat down to talk, she revealed to me: "I don't understand why I have to get these awful treatments, since my doctor told me my cancer was gone." I asked her if she had asked her doctor that question and, of course, she had not, being from the old school where the doctor is never questioned. We did a role-play until she was comfortable enough to talk to the doctor after our meeting. Instead of being angry, he was more than happy to sit down and discuss her question. He explained that women with her type of cancer were less likely to relapse if they got six months of chemotherapy. That's all it took for Millie to revert back to her old self and to be a pleasure again coming back to the clinic. In other words, a little information can go a long way to understanding.

9. **Cancer is a social disease. Improving the quality of relationships can become a source of healing support for me. This also includes my spiritual relationships.**

Betty was a late-fifties breast cancer patient. She taught me that going through a traumatic ordeal like cancer can make you strong enough and confident enough to stand up for yourself and ask for what you deserve. After a course of individual psychotherapy and regular attendance at several of the support groups she learned assertiveness and improved her self-esteem enough to confront her husband and develop a more balanced relationship. She finally got the dog she had always wanted, even though it took a bald head to get it. She was also pleased that her husband finally accepted Jesus as his Savior, which helped them grow even closer.

10. **Acceptance is not giving in. As Norman Cousins said, "Don't deny the diagnosis. Try to defy the verdict." Keep in mind, statistics apply to groups, not to a given individual. At least one person has defeated each type of cancer.**

Bob was a mid-sixties lung cancer patient who was given only three to six months to live after finding he had metastatic disease. He taught me that miracles happen and that it is important to accept the disease and move on to what is healthy for an individual. His medical team, especially his radiation oncologist, had not been encouraging by informing him that nothing would help. After a great deal of depression, he came to the support group and gained inspiration by seeing other cancer patients who were not just surviving but thriving. He cobbled together a program involving a vegetarian diet, vitamins and other supplements, regular meditation and visualization, exercise, support groups, and church attendance. This was more than eight years ago. He has become our "Poster Boy."

11. There is always hope, but what I hope for may change over time. As Dr. Carl Simonton said, "In the face of uncertainty, there is nothing wrong with hope."

Sharon was an early-forties lymphoma patient who was told to put her affairs in order after she had developed a recurrence. Rather than think that would be her last Christmas, she decided to get a second opinion. She had developed a tumor in her chest that made it increasingly difficult to breath. She had radiation, but the tumor did not appear to decrease. No one would operate on her in Kalamazoo, so she went to the Mayo Clinic. They were willing to operate and found that the tumor was cancer but had been so large that when the radiation killed it, it was too large for the body to get rid of it on its own. She began to recover and decided she needed to make some changes. She quit her high stress job and moved up North to be with her family. She got a different job, a new boyfriend. Sharon also started a new cancer support group.

12. My personal dignity and quality of life are always the best measures of success. It takes only a split second to die and all the rest of my time goes to living as best and joyfully as possible. Death is not a failure.

Clancy was a teenage leukemia patient who had set the record for re-admissions to the pediatric floor over her years of struggle with that disease. She was a real favorite of the nurses and they always expected her to bounce back. So did her mother, so it was hard to believe she was dying this time. While others were afraid to face things, she was wise beyond her years and confided in me that she didn't think she would make it to December so she wanted Christmas in August. It was a major celebration with a miniature Christmas tree, present exchange, and memory swap. She choreographed her death and died satisfied that she had lived every minute as best she could. She taught us that you don't have to live many years to touch many people deeply. She taught family, friends, me, as well as the rest of the staff, that you can't cheat Death, but you can give him a run for his money.

For many people, being diagnosed with cancer is like being dropped into a combat zone where everyone is speaking a foreign language. Whether the diagnosis is a sudden surprise or a suspicious symptom takes weeks to diagnose, it can be overwhelming and leave one feeling powerless and out of control. Cancer and its treatment bring a great deal of emotional and physical stress. Because stress can further complicate treatment and compromise the body's defenses, managing stress is a valuable way to fight back and regain a sense of control.

This manual is designed to help cancer patients and their loved ones become empowered by learning more effective ways of managing stress and igniting their "fighting spirit." Steven Greer and his colleagues ("Psychological Response to Breast Cancer and 15-Year Outcome," Lancet, 1990, i, 49-50) studied breast cancer patients. He measured their personalities and found that a greater number of the women who had "fighting spirit" survived than women who were "deniers," "stoic acceptors," and "pessimists." While this remains controversial, other researchers like Carl Simonton and Stephanie Matthews-Simonton, David Spiegel, and Fawzy Fawzy have found that attitude may help improve survival. In order to spark our "will to live," we must first clearly identify the important and joyous things for which we want to live. See the following story: "What are the Big Rocks in your Life?"

As traumatic as cancer can be, it can also serve as an opportunity to evaluate life and re-establish priorities. During the decades of my clinical practice, I have had many people identify "gifts" they have received by having cancer. It has caused them to "take an inventory" of their priorities. In your own journey, if you are ready to explore positive changes, then begin by doing some goal setting. Look at both short and long-term goals. Also look at goals in many different areas of your life. Below are some hints to help you structure goals that will help direct your energy and increase your motivation. Remember two cardinal rules:

❶ **Gear for success by staying positive.**
"Optimism is a ninety-year-old couple looking for a house—near a school."
—Dr. Murray Banks

❷ **No step toward your goal is too small.**
"If you are going to try cross-country skiing, start with a small country."
—*Saturday Night Live*

What are the big R OC K S in your life?

An expert in time management was speaking to a group of business students. As the man stood in front of this group of high-powered overachievers, he said, "Okay, it's time for a quiz." Then he pulled out a one-gallon, wide-mouthed mason jar and set it on a table in front of him. Then, he produced about a dozen fist-sized rocks and carefully placed them, one at a time in the jar. When the jar was filled to the top and no more rocks would fit inside, he asked, "Is this jar full?"

Everyone in the class said, "Yes."

Then he said, "Really?" He reached under the table and pulled out a bucket of gravel. Then he dumped some gravel in and shook the jar causing pieces of gravel to work themselves down into the spaces between the big rocks. Then he smiled and asked the group once more, "Is this jar full?"

By this time, the class was onto him. "Probably not," one of them answered. "Good!" he replied. And he reached under the table and brought out a bucket of sand. He started dumping sand in and it went into all the spaces left between the rocks and the gravel. Once more he asked the question, "Is this jar full?"

"No!" the class shouted. Once again he said, "Good!" Then he grabbed a pitcher of water and began to pour it in until the jar was filled to the brim. Then he looked up at the class and asked, "What is the point of this illustration?" One eager beaver raised his hand and said, "The point is, no matter how full your schedule is, if you try really hard, you can always fit some more things into it!"

"No," the speaker replied, "that's not the point. The truth this illustration teaches us: If you don't put the big rocks in first, you'll never get them all in at all."

(Internet source unknown)

SHORT-TERM GOALS

We need goals, no matter how small, to get us out of bed in the morning. Goals help us focus and direct our energy. Goals keep us motivated. One patient who had multiple, chronic health problems once told me the only thing that kept her going was curiosity to see what was in the news. While you may need more than that, the point is we need something that generates our life energy. Motivation and successful goal setting are not necessarily inborn, especially when you feel sick or your energy is zapped by cancer treatments. Think of short-term goals as daily, weekly, or monthly accomplishments. The following *Dos and Don'ts* offer some helpful guidelines:

Do set goals as a way of taking the first step toward making them happen.

Example: Make a definite plan to get together with friends this weekend. Don't hesitate to commit because you have your chemotherapy on Tuesday and are afraid you may feel badly Saturday. If you do feel badly, you can always cancel. Another option would be to have friends come over and watch a video or play cards, so if you began to feel tired, you could retreat to your room. Again, making the plans is the first step. Good friends will understand and be flexible should things need to change.

Don't stop setting goals because you are sick or might not feel well.

If you have this attitude, your world could begin to shrink dramatically.

Do set goals first, then begin to direct your energy and put yourself in the mindset to make things happen.

Setting realistic goals will help you focus and taking tangible steps will help you regain some sense of control. Pick something specific, concrete, and small to start. People often think of goals as something major that takes a great deal of time to accomplish. Long-term goals may be that way, but you need to break those goals into smaller steps, otherwise you might deflate your motivation. Example: I want to be strong enough to go back to work full-time. That may start with doing small jobs of twenty to thirty minutes at home.

Don't get bogged down in feeling bad about how much you were able to do before.

You many need to grieve your losses, but don't get angry with yourself because you can't do what you did before. Be supportive, not judgmental. Eliminate "shoulds" and "shouldn'ts" from your vocabulary.

Do focus on where you are now as a starting point to build back up.

You may be able to do a great deal more than you expect. You may have to approach things differently now. Example: you may have been able to read for hours straight before your treatments started. Now you might find your attention wanders more easily. You might have to read in fifteen-minute blocks and take a brief break. You may be frustrated at first, but you may begin to gradually build up concentration between breaks.

It is not cheating to keep your goals small and easily obtainable. We are encouraged and motivated more by success than failure. You can always increase or decrease goals as you progress. Avoid setting arbitrary goals because you think you should be more fully recovered than you are.

Do keep records.

Keeping graphs or charts may motivate you. You can literally see your progress and know exactly if you are sticking to your goals or if adjustments have to be made.

Do enlist social support.

Let your family know what you need help with and what you can do yourself. Let them know specifically what you may need from them. Example: I want to do the laundry, but I need you to carry it downstairs to the washer.

Do pick one to three goals for each day.

Set your goals after you have awakened and had a chance to assess how you are feeling physically and emotionally. What are you capable of doing? What would nurture you the most? If you feel really low, then pick goals that require less physical effort or energy (e.g. reading, journaling, gentle crafts). If you are feeling fairly well, adjust your sights accordingly (e.g. gardening, cleaning, exercise). If you are feeling depressed, you may need a good laugh or enjoy doing something sensual. You may need to explore what is making you sad and identify something to do about it. No matter how small you may feel they are, still set up to three goals. Do not write out a laundry list. Long lists are overwhelming and may short-circuit your motivation. Instead, feel good about what you've done and if you feel good and do more, consider that a bonus.

While short-term goals may be valuable in and of themselves, they should tie into long-term goals, which are related to the high priorities of our lives.

LONG-TERM GOALS

Day-to-day goals are important and short-term goals may be thought of as things to achieve in the next one to three months. The process of goal-setting keeps our immediate focus. Long-term goals reflect our major values and what is truly important in life. They guide our short-term goals. Example: If our long-term goal is to have a healthier diet, the first step may be the short-term goal of eating less than 30 fat grams a day. Another example: I value my family, so I will try to resolve differences with my son by doing one mutually agreeable activity a week.

Often when individuals are diagnosed with cancer, they may feel discouraged from setting long-term goals. You may feel that you will not be alive to enjoy reaching the goal. Resist that with all your heart, until your heart is no longer into living. Continue to plan for the future, because that is the first step to getting there! Even plan your funeral and life celebration, if that will ease your mind and free up energy for living. One of my patients who had metastatic lung cancer was told he had three to six months to live. He was initially reluctant to plan, but began a variety of lifestyle changes and has enjoyed a number of trips in the six years he has remained alive since his diagnosis.

Try to avoid the mindset: "I'm not sure if I'll feel up to it." This begins to constrict your world. Instead, make your plans for travelling or for your child's graduation or a family reunion next summer. Planning gives you something to look forward to and sets you on course to achieve it. Of course, if you do not feel well when the event comes around, you should cancel. But by planning, it is more likely that you will end up doing it. Otherwise, your world begins to shrink and you become even less confident to make plans. There is also the enjoyment of the planning itself. As we used to say in the '60s, "The journey is part of the trip."

When making long-term goals, keep in mind the one to five year range. Again that may seem unreachable, but that's where it starts. Make your goals concrete and specific. Break them into smaller steps that are more easily reachable. Set goals in the different areas of life such as family (I will write a card to a cousin with whom I was close...), recreational (I will learn a new hobby like oil painting that I've always wanted to try...), and social, as well as vocational, health and spiritual/religious. Actually write out your goals. Looking at them makes them more real and may increase your commitment to making them happen. You may use the format on the following pages or devise your own personal system for listing short and long-term goals.

HEALTH

Short-term goals:

1) _____

2) _____

3) _____

Long-term goals:

1) _____

2) _____

3) _____

FAMILY

Short-term goals:

1) _____

2) _____

3) _____

Long-term goals:

1) _____

2) _____

3) _____

SOCIAL

Short-term goals:

1) _____

2) _____

3) _____

Long-term goals:

1) _____

2) _____

3) _____

VOCATIONAL

Short-term goals:

1) _____

2) _____

3) _____

Long-term goals:

1) _____

2) _____

3) _____

RECREATIONAL

Short-term goals:

1) _____

2) _____

3) _____

Long-term goals:

1) _____

2) _____

3) _____

SPIRITUAL

Short-term goals:

1) _____

2) _____

3) _____

Long-term goals:

1) _____

2) _____

3) _____

PRACTICAL MATTERS

Short-term goals:

1) _____

2) _____

3) _____

Long-term goals:

1) _____

2) _____

3) _____

Basic Information

on the

Stress Response

AN OVERVIEW OF THE STRESS RESPONSE

Stress is the complex pattern of physical, cognitive (thinking)-emotional, and behavioral responses **to meet the demands placed on us**. These three levels interact and can increase our reactions to stress, unless we take conscious steps to change them. In order to achieve really noticeable stress management, we need to impact all three areas.

THE THREE LEVELS OF THE STRESS RESPONSE

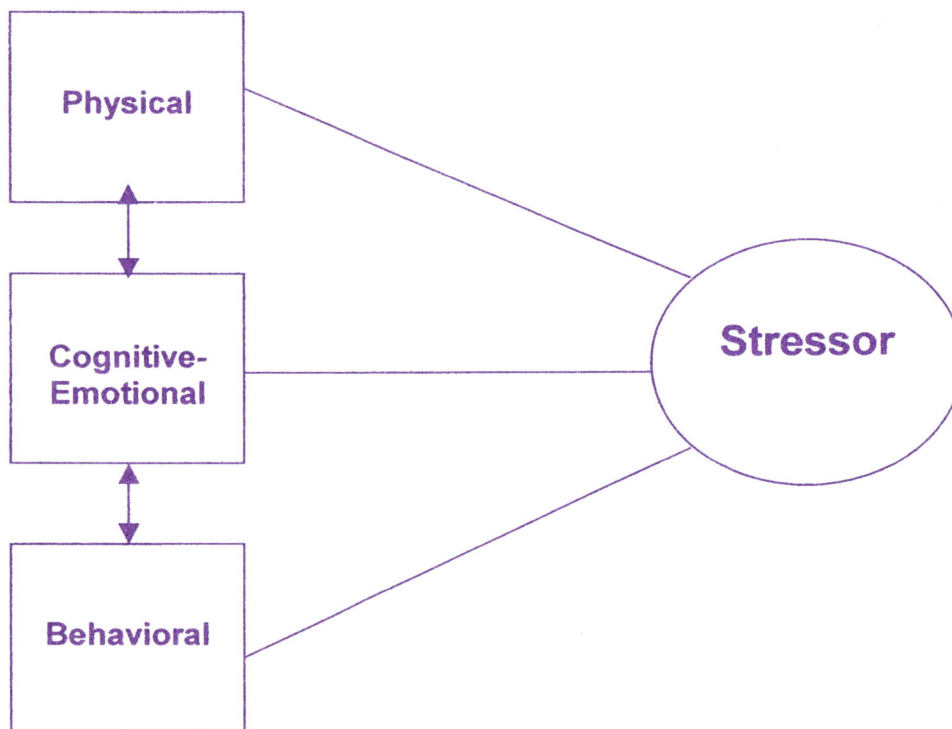

Although the Stress (Fight-or-Flight) Response is innate, our pattern becomes learned or conditioned over time. Therefore, every time we are exposed to **a situation that we evaluate as stressful**, our same response pattern becomes triggered. Our stress response then becomes more of a habit, making it even more conditioned and automatic or unconscious. Stress then seems beyond our control. The first step in understanding how stress affects you is to examine your pattern of responses. What physical changes (sensations) do you notice? What kinds of things do you say to yourself? What kinds of actions do you take? Then you can begin to change the particular factors that are producing stress.

THE NATURE OF THE STRESS RESPONSE

The true nature of the stress response is to help put our minds and bodies in the best state to deal with true threat (stress). If you look at what nature intended, the fight-or-flight response is amazingly adaptive for us. This is great! If we perceive a threat and deal with it immediately and directly (mentally or physically), then no real damage occurs. That is just normal wear and tear. It looks something like this:

Normal Stress Response
——————————————➤

There are three conditions where stress does become damaging: when we stay in our stress response for too long a time, when we react with our stress response too frequently, and when we have too extreme a response. Those reactions look something like these:

Prolonged Stress Response
——————————————➤

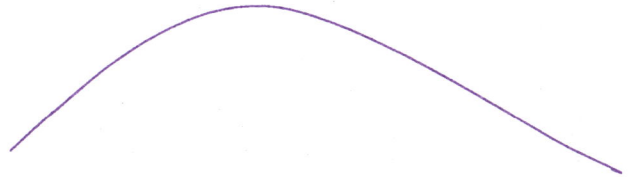

Too Frequent Stress Response
——————————————➤

Too Extreme Stress Response
——————————————➤

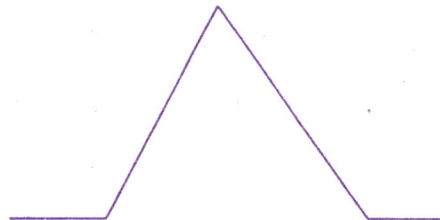

So when we are talking about stress management, we want to reduce the amount of time spent in the stress response, react with the stress response less often especially when we aren't threatened, and contain our response. We cannot eliminate stress, because stress occurs in every change situation.

THE ONLY THING THAT DOESN'T CHANGE
IS THAT THINGS KEEP CHANGING!

Some research indicates that some stress or arousal (as we psychologists prefer) may even be healthy. To a certain extent, arousal is important for our best performance regardless of the activity we are performing. The relationship between stress (arousal) and performance looks like this:

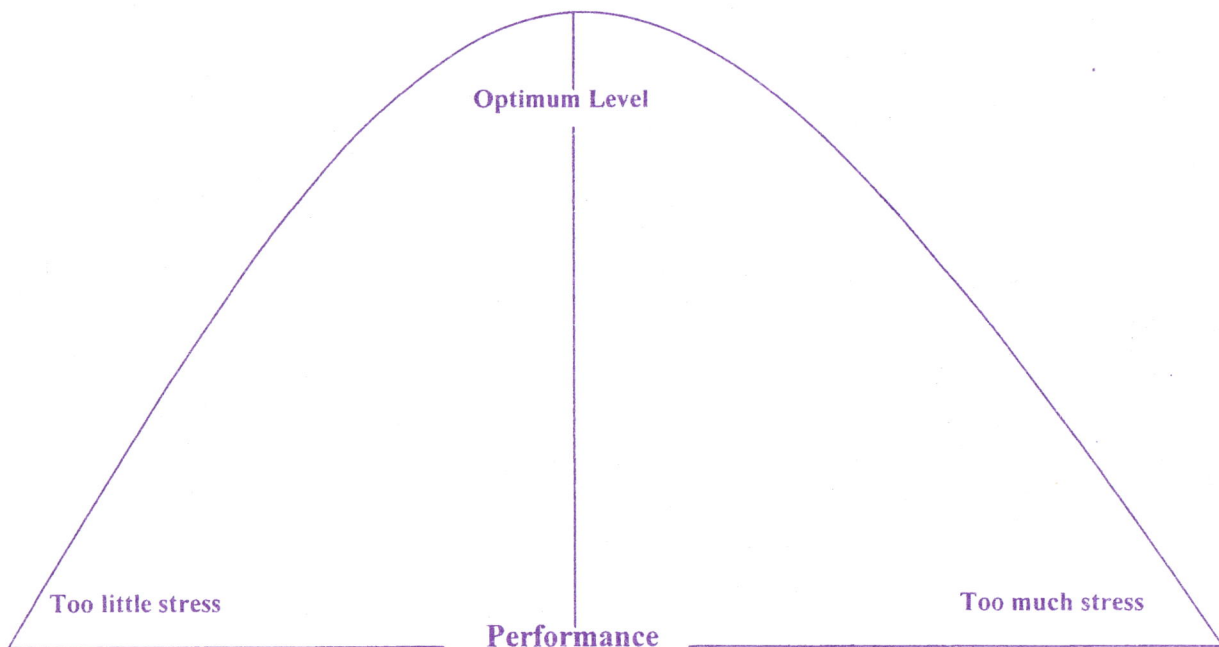

Optimum Level

Too little stress   **Performance**   **Too much stress**

When we have too little arousal, our performance is flat, lackadaisical, and inattentive. When we have too much arousal, our performance is tense, scattered, and hyper-vigilant. We are trying to manage stress so that we remain attentive, directed, and focused without any physical or mental discomfort. Our stress response is systemic, which means it can affect any and all of the body's systems. If we don't pay proper attention to our stress, it can cause real emotional and physical problems. Some experts believe as much as 80 percent of all doctor visits or health problems may be caused or aggravated by an exaggerated stress response. Some of the adverse reactions to stress are outlined in the following chart.

SOME ADVERSE STRESS REACTIONS

PHYSICAL CHANGES

- Altered breathing
- Increased heart beat
- Increased blood pressure
- Cholesterol changes
- Altered blood flow
- Increased blood sugar
- Increased muscle tension
- Increased stomach acid
- Increased sweat gland activity
- Hormonal changes
- Altered and decreased immunity

COGNITIVE-EMOTIONAL CHANGES

- Impaired problem solving
- Shift in attitude toward self and others
- Faulty perception of events
- Confusion in values and expectations
- Feeling overwhelmed and overly sensitive
- Feeling angry, irritated, frustrated, or resentful
- Feeling panicky or overly anxious
- Feeling depressed, apathetic, hopeless, lethargic, or unmotivated

BEHAVIORAL CHANGES

- Insomnia
- Loss of appetite or increased eating
- Conflict
- Crying spells
- Drinking, drug-taking, smoking
- Avoidance and/or procrastination
- Withdrawal and social isolation
- Ineffective or inefficient behavior – burnout
- Suicide attempts
- Sickness

THE ADDITIVE NATURE OF STRESS

Stress tends to build up when we ignore our internal signals, until our bodies let us know what is happening. When our head feels like it is about to explode, or our neck is stiff and sore, or our stomach is in a knot, we finally notice that our stress has accumulated. While this may feel as though it has occurred instantly, often our tension has been building up, although it may be subtle enough that we feel "normal." This typically happens because we are focused outwardly and thinking about what we have to do. We are not generally very conscious of what our body feels like until our discomfort breaks through our conscious awareness.

To better deal with stress, we need to begin paying more attention to our bodies. We need to do this for one simple reason: YOUR BODY NEVER LIES! Our true, and often subconscious, feelings are in one's body and one's is like a child. If it doesn't get our attention with a small symptom, it will keep building until it becomes a major symptom that can no longer be ignored. The more we become conscious of how our body feels, the more likely we will head stress off at the pass. In other words, if we can pick up on our stress when it is first beginning to rise, then a brief intervention may be all we need. A couple of minutes of focused breathing, changing our internal self-talk to something more calm and reassuring, or doing some simple stretches may be enough to relax us a notch or two and keep our stress under control. Instead, what typically happens is that we ignore our stress and keep pushing until it is very high, and then it is more difficult to calm down. This often makes us feel like our stress is out of control, because it is.

This escalating threshold is illustrated on the following graph. The short arrow represents our awareness of stress as it first begins to rise. That is where we need to try something immediate and brief to effectively manage our stress. The longer arrow shows where we usually notice our stress, when it breaks through our threshold of conscious awareness. Intervention to manage stress may still work, but it takes much more effort and we often become frustrated and stop before we successfully reduce our stress. We may then start reinforcing an inaccurate conclusion that our stress is out of control and there is nothing we can do about it. The common sense approach is to begin noticing where our stress level is throughout the day and if it is increasing, we can start intervening earlier. This approach is also called "mindfulness" and has been explained well by Jon Kabat-Zinn in his books, *Full Catastrophe Living* and *Wherever You Go There You Are*. It is a similar idea to **The Feeling Pause** exercise that will be discussed later in this manual.

CONSCIOUS THRESHOLD

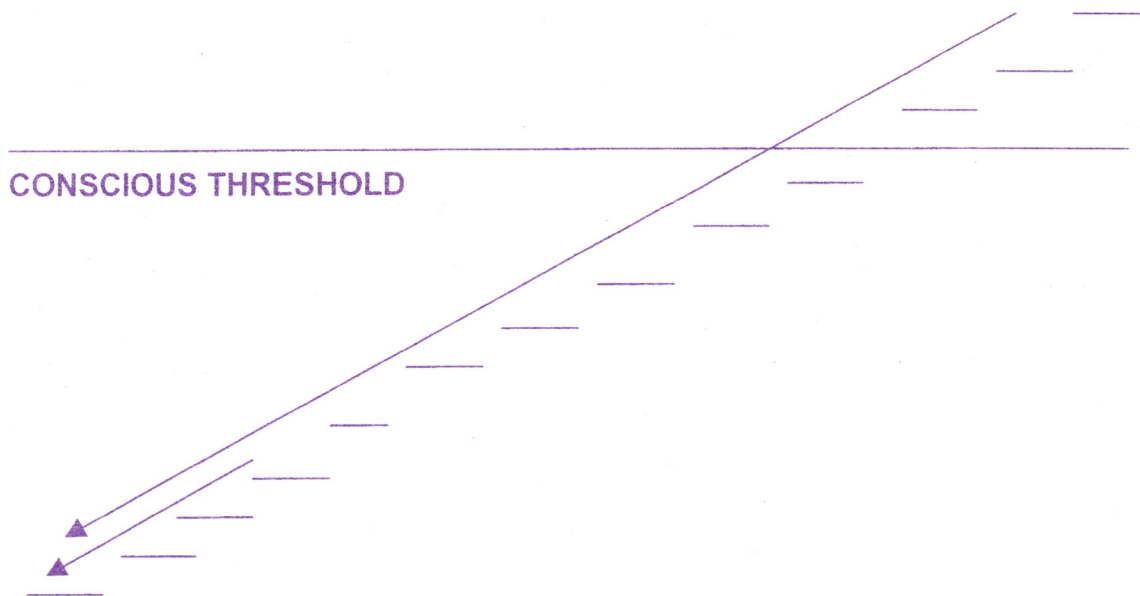

This shows the additive nature of stress and how early intervention throughout the day helps control stress more easily. The short lines represent hassles or stressors that we face. With each one can come additional physical and/or emotional tension, especially if we are not paying conscious attention. If enough stress is added, then it will finally break through our conscious threshold often in the form of physical symptoms like a headache, neck stiffness, or stomach problems. If we learn to monitor and identify our stress earlier (the shorter arrow), then it will be easier to reduce our stress by taking a few deep breaths, imagining a pleasant scene, doing some muscle stretching, or talking to ourselves in a calming way. It is much more difficult to intervene when our stress has accumulated (the longer arrow).

The additive nature of stress also comes into play when we face a situation that makes us nervous, such as a doctor appointment or a chemotherapy treatment. Our stress may start increasing days or hours before we come to the clinic. Apprehension, even when it is subconscious, builds until it peaks in the situation. There our anxiety level stays high for up to forty to sixty minutes before it starts to fall. If each step of the way, we can substitute the relaxation response for the stress response (like Systematic Desensitization which is discussed later) we can control our stress much better.

THE MIND—BODY—SPIRIT CONNECTION
A Holistic View of Health

Sit in a comfortable position as you begin to read this page. After you have read this paragraph, close your eyes and imagine yourself in the kitchen of your home. Walk to the refrigerator, open the door, and pull out a bright yellow lemon. Notice the cold, slightly bumpy texture of the lemon's skin as you take it over to the cutting board. Notice how firm and juicy the lemon is as you slice off a piece. Pick up the slice of lemon and place it in your mouth. Now take a big bite and notice how your body responds.

Depending on how involved you were with the mental scene, you may have noticed that your mouth began watering. This is an example of how interwoven the mind and body truly are. Dr George F. Solomon, a University of California psychiatrist along with Rudolf H. Moos first speculated a link between immunity and emotions in 1964. In an article called "The Emerging Field of Psychoneuro-immunology With a Special Note on Aids," Solomon (*Advances*, Vol.2, No.1, Winter, 1985, 6-19) suggested that the mind and body are inseparable and he described a number of principles that hypothesized how the brain and central nervous system influence all sorts of physical processes. Solomon has expanded his list of postulates over the last fifteen years and research has validated much of it.

Current research results have caused us to reassess our view of health and the greater involvement our mind, brain, and spirit — thoughts, attitudes, personality, beliefs and ability or inability to handle stress — have on our overall health. Likewise, this research has not ignored the vital role our physical bodies play in our psychological health. It is breaking down the notion that the mind and body are separate and renders Woody Allen's question irrelevant. ("Is there a mind/body problem? And which is better to have?")

The concept of illness is truly psychosomatic. Many estimates have attributed 50 to 90 percent of all disease and doctor office visits to emotional or stress-related conditions. This does not mean that headaches or anxiety are imaginary or reside strictly in our minds, as most people have come to believe. The term "psychosomatic" refers to both mind (psycho) and body (soma) and the inter-relation between them in the disease process. Any illness arises out of a complexity of factors including genetic, physical, emotional, social, dietary, environmental, and spiritual forces.

With any given individual, as well as ailment, there may be a greater or lesser involvement from any of these dimensions of the person's life. To use migraine headaches as an example, we know that there are dozens of physical events happening, which include changes in the blood vessels and numerous biochemical and electrical disturbances. There may be a genetic predisposition to these physical changes. These physical changes may be triggered by a wide variety of events, spanning social and emotional, as well as environmental factors. Headaches may be caused by a family argument or work deadline. Feeling guilty about our behavior, not attending church, or avoiding a traumatic event, may also lead to a headache. Environmental irritants, bad lighting, certain foods that are high in tyramine (an amino acid), missing meals, awkward posture, and hormonal response may also provoke headaches. Translated, physical or psychological stress can throw our mind-body-spirit system out of balance.

The same is true of cancer. Cancer is a complex disease (actually it is between 100-200 different diseases!) and is not just related to one specific factor. Some people are at risk for cancer from a genetic standpoint and may carry oncogenes (genes that when expressed, facilitate cancer) or have problems with tumor suppression gene (genes that when expressed, make cancer more unlikely to occur). There may be a physical factor, such as chronic irritation by acidic reflux in the esophagus or excessive hormones, as in breast or prostate cancer. Stress overload or an unassertive personality may decrease our bodies' defenses and allow cancer to be harder to detect. The same may be true of people who feel they have lost their meaning in life, are depressed, or lack a spiritual connection. A high fat diet and diets lacking in fruits and vegetables have been linked to colon cancer and ultraviolet radiation, not to mention cigarette smoke, which are potent carcinogens. What makes it even more confusing is that two people with the same cancer may have greater or lesser involvement from the same risk factors. Unfortunately, there is no simple equation with designated weights for all the factors that can tell us the exact risk someone has for contracting cancer.

Stress May Aggravate Nearly Every Illness

We are approaching a clearer understanding of how stress affects us and may even make us more prone to illness. Recent research supports the notion that stress may partially cause or aggravate nearly every human ailment. Ulcers used to be the first association that came to mind when one thought of a stress-related problem. Although the primary factor for stomach ulceration is a bacterium, some people do secrete more stomach acid when stressed and get ulcers. Stress plays a significant role in other gastrointestinal problems such as irritable bowel syndrome, spastic colon, irregular or rapid heart rate, high blood pressure, chronic pain, migraine or tension headache, anxiety disorders such as panic attacks, depression, as well as alcohol and other substance abuse. To a lesser degree stress may contribute to aggravating diabetes, asthma, and seizure disorders.

The latest research is also illustrating how stress can affect our body's natural defense system. Cohen and some of his colleagues ("Psychological Stress in Humans and Susceptibility to the Common Cold," *New England Journal of Medicine*, 1991, 325, 606-612) have discovered that under controlled exposure to cold viruses that significantly more high-stressed individuals actually developed cold symptoms than low-stressed individuals. Other researchers have found that many forms of stress may have a detrimental impact on both quality and quantity of different types of white blood cells in our body's defenses. These results may relate to why we are more likely to catch colds or be more susceptible to viruses when we have been under large amounts of stress. Think back over your history and see if that has been your experience.

If stress depletes our body's defenses, it may be involved with developing cancer. First, it is important to state that stress does not cause cancer. It does not act like a carcinogen (a cancer causing substance), causing the DNA of cells to become damaged, as substances like radon or cigarette smoke have been clearly documented to do. But stress has been shown to slow DNA repair and would-be healing. More importantly, however, stress has been shown to negatively affect our Natural Killer (NK) cells, which are a type of white blood cell that spontaneously kills abnormal cells, such as cancer cells. Impaired NK cells may make it easier for cancer to go undetected or recur. Think back to the time before your diagnosis. Were you under a great deal of stress? Had you lost someone close to you or a job that was really meaningful? Check the list of major life stressors on the Social Readjustment Rating Scale (SRRS) developed by Holmes and Rahe in the 1960s to see if there may have been a great deal of stress twelve to eighteen months before your cancer was diagnosed.

Now again take a moment to close your eyes and imagine yourself in a recent situation of high anxiety you've experienced. Maybe it was a conflict with someone in your family, or someone at work, or perhaps frustration from failure to achieve a goal.

Notice the impact of these thoughts and images on your body. What happened to your heartbeat and your breathing rate? What about the temperature of your hands and feet? Did you clench your teeth or your fists? How tight were your neck and shoulders? Did your stomach growl?

For years stress researchers have tried to unravel the nature of the stress response and its health effects. The late Dr. Hans Selye, the foremost researcher of the stress response, proposed a theory about stress known as the **General Adaptation Syndrome** (GAS). Stress is the body's response to meet a perceived threat. In prehistoric times, that threat was largely physical. Now it is usually a psychological or social threat. This model states that any of our responses to any kind of stress would show the same basic kind of physical responses, such as increased heart beat and breathing rate, change of blood flow away from the hands and feet to the large muscles of the body, increased blood sugar levels, and muscle bracing. Further, even though we would all generally react in the same manner, we each have our own individual differences due to our genetic make-up and unique learning history. This would account for one person getting stomach distress as opposed to muscle contraction headaches. The next time you find yourself under stress, notice what is happening to your body.

Fight-or-Flight Response

Dr. Selye and many stress researchers contend that stress itself is not necessarily damaging. When we look at how the stress response evolved, we can understand that it operates and functions as a wonderfully adaptive response to meet the demands placed on us. If we look at a primitive cave person happening upon a saber-toothed tiger, the fight-or-flight response (as the stress response is also called) kicks in. The stress response would enable our cave person to fight the saber-toothed tiger and kill it, or get the heck out of there. While we do not encounter too many saber-toothed tigers, we may encounter a mugger or a life-threatening situation, such as an auto accident or a cancer diagnosis. The stress in our lives has shifted more to psychological or social stress arenas. We stay angry thinking about a goal we failed to achieve or we worry about what the doctor will say at our next appointment. Our bodies, however, do not distinguish these threats. If we have not learned to manage our stress response, we respond as though we had encountered a saber-toothed tiger.

Today's Stressors Remain Unresolved

What happened in primitive times is that the fight-or-flight response was triggered, the stressor was a distinct physical threat, which was dealt with, and the body returned to a state of balance with a minimal amount of distress. Today, since our stressors are not as distinct, we often do not get a chance to resolve them quickly. When our bodies constantly initiate the stress response, the wear-and-tear may result in any of the stress-related health care problems mentioned earlier.

Researchers are discovering that it is *not the particular situation* that produces stress as much as the person's perception and attitude about the event, and his or her ability to adapt to the necessary change. In 1967, Richard R. Rahe and Thomas H. Holmes (**The Social Readjustment Rating Scale**, *Journal of Psychosomatic Research, 11, 213-218*) developed a scale of major life stressors, which assigned certain point values to these major life events (i.e., Death of a spouse = 100 points; Divorce = 73 points, etc.). They discovered that people who were hospitalized were more likely to have had a large amount of stress twelve months prior to their hospitalization. They predicted that a person scoring over 300 points had a 90 percent chance of having an adverse change in health. This instrument is included next. Being diagnosed with cancer brings a number of unique stresses. A special awareness tool developed for cancer patients (**Stressful Situations Survey for Cancer Patients (SSSCP)**) by this author also is included. This instrument can help you identify situations where you could apply the techniques in this manual.

Researchers have not found the correlation between illness and major life stresses to be as close as first suspected. Hassles, those small irritants that may occur frequently add up, and may be as detrimental as major stresses. If we have no major life stresses, but are continually irritated by other people driving too slowly or pressured to be endlessly productive, we may be putting ourselves under a great deal of needless stress that can exert a tremendously damaging effect on our bodies. Instruments have been devised to assess the detrimental effects of hassles, too.

If we were to examine the vast majority of stressors that we face, we would see that our thoughts and attitudes about the events determine whether they are stressful or not, and to what degree.

SOCIAL READJUSTMENT RATING SCALE (SRRS)
Developed by Thomas H. Holmes and Richard R. Rahe (1967)

Directions:
Mark those experiences that apply from the last twelve months or that you anticipate in the next twelve months. Mark the number of points that apply and add them up.

Past 12 mos.	Future 12 mos.		Points
_____	_____	1. Death of spouse	_____ 100
_____	_____	2. Divorce	_____ 73
_____	_____	3. Marital separation	_____ 65
_____	_____	4. Jail term	_____ 63
_____	_____	5. Death of close family member	_____ 63
_____	_____	6. Personal injury or illness	_____ 53
_____	_____	7. Marriage	_____ 50
_____	_____	8. Fired from job	_____ 47
_____	_____	9. Marital reconciliation	_____ 45
_____	_____	10. Retirement	_____ 45
_____	_____	11. Change of health of family member	_____ 44
_____	_____	12. Pregnancy	_____ 40
_____	_____	13. Sex difficulties	_____ 39

_____	_____	14. Gain of new family member	_____	39
_____	_____	15. Business readjustment	_____	39
_____	_____	16. Change in financial state	_____	38
_____	_____	17. Death of a close friend	_____	37
_____	_____	18. Change to different line of work	_____	36
_____	_____	19. Change in number of arguments with spouse	_____	35
_____	_____	20. Mortgage over $75,000	_____	31
_____	_____	21. Foreclosure of mortgage or loan	_____	30
_____	_____	22. Change in responsibilities at work	_____	29
_____	_____	23. Son or daughter leaving home	_____	29
_____	_____	24. Trouble with in-laws	_____	29
_____	_____	25. Outstanding personal achievement	_____	28
_____	_____	26. Wife begins or stops work	_____	26
_____	_____	27. Begin or end school	_____	26
_____	_____	28. Change in living conditions	_____	25
_____	_____	29. Revision of personal habits	_____	24
_____	_____	30. Trouble with boss	_____	23
_____	_____	31. Change in work hours or conditions	_____	20
_____	_____	32. Change in residence	_____	20
_____	_____	33. Change in schools	_____	20
_____	_____	34. Change in recreation	_____	19
_____	_____	35. Change in church activities	_____	19
_____	_____	36. Change in social activities	_____	18
_____	_____	37. Mortgage or loan less than $10,000	_____	17
_____	_____	38. Change in sleeping habits	_____	16
_____	_____	39. Change in number of family get-togethers	_____	15

_____ _____	40. Change in eating habits	_____ 15
_____ _____	41. Vacation	_____ 13
_____ _____	42. Christmas	_____ 12
_____ _____	43. Minor violations of the law	_____ 11

TOTAL POINTS: _____

STRESSFUL SITUATIONS SURVEY FOR CANCER PATIENTS (SSSCP)

Morry Edwards, Ph.D.

The experience of cancer may bring about situations that people find stressful. Different people may find certain situations more stressful than others. This survey provides an opportunity for you to assess which situations are most uncomfortable for you as the first step toward reducing that stress.

Directions:

Beside each situation, circle the discomfort you feel on a scale of 0 to 3, with 0 being no noticeable discomfort, and 3 being extremely severe stress or discomfort. Next to the scale, write down approximately how often that situation occurs in an average month for you. If it does not occur, then mark NA for not applicable. Lastly, under each question is room for you to list situations that you find stressful, but which have not been included.

EXAMPLE:

17. Waiting in the waiting room for a   0   1   2   3
chemotherapy treatment

Name_____ Age_____ Sex _____

Diagnosis_____ Today's Date _____

Time since diagnosis_____ Years_____ Months_____

Treatments (scheduled or finished)_____
_____
_____
_____

Medical Care for Doctor's Appointments

1.	Coming to the clinic for an examination	0	1	2	3
2.	Arriving at the clinic late	0	1	2	3
3.	Waiting in the waiting room before your appointment	0	1	2	3
4.	Waiting in the examination room before the doctor comes in	0	1	2	3
5.	Doctor being late for your appointment	0	1	2	3
6.	Asking your doctor questions about treatment	0	1	2	3
7.	Asking your nurse about treatment	0	1	2	3
8.	Having blood drawn for your lab tests	0	1	2	3
9.	Having an X-ray	0	1	2	3
10.	Having a medical test, such as a bone marrow	0	1	2	3
11.	Not receiving clear information	0	1	2	3
12.	Difficulty communicating with medical team	0	1	2	3
13.	_____	0	1	2	3
14.	_____	0	1	2	3
15.	_____	0	1	2	3

Treatments

16.	Going to the clinic for a chemotherapy treatment	0	1	2	3
17.	Waiting in the waiting room for a chemotherapy treatment	0	1	2	3
18.	Hearing that your lab tests are satisfactory for treatment	0	1	2	3
19.	Hearing that your lab tests are not satisfactory and you will not be treated	0	1	2	3
20.	Starting the infusion of the chemotherapy	0	1	2	3
21.	Finishing one course of chemotherapy treatment	0	1	2	3
22.	Completing your entire chemotherapy treatment	0	1	2	3
23.	Going for a radiation therapy treatment	0	1	2	3
24.	Waiting in the waiting room for a radiation therapy treatment	0	1	2	3
25.	Receiving a radiation therapy treatment	0	1	2	3
26.	Finishing radiation therapy treatment	0	1	2	3
27.	Completing your entire radiation therapy	0	1	2	3
28.	_____	0	1	2	3
29.	_____	0	1	2	3
30.	_____	0	1	2	3

Hospitalization

31.	Going into the hospital for routine treatment	0	1	2	3
32.	Going into the hospital for planned surgery	0	1	2	3
33.	Going into the hospital with an unexplained problem	0	1	2	3
34.	Sharing a room with a stranger	0	1	2	3

35.	Having to follow hospital routine	0	1	2	3
36.	_____	0	1	2	3
37.	_____	0	1	2	3
38.	_____	0	1	2	3

Personal and Social

39.	A change in your physical appearance	0	1	2	3
40.	Seeing friends that you have not seen in a while	0	1	2	3
41.	Having friends visit longer than you would like	0	1	2	3
42.	Feeling abandoned by friends/family	0	1	2	3
43.	A decrease in activity level	0	1	2	3
44.	Difficulty with physical activities	0	1	2	3
45.	The cost of your medical treatment	0	1	2	3
46.	Loss of affection	0	1	2	3
47.	A change in sexual activity	0	1	2	3
48.	Talking about your true feelings with family members	0	1	2	3
49.	Talking about your true feelings with friends outside the family	0	1	2	3
50.	Feeling overprotected or smothered by family/friends	0	1	2	3
51.	Lack of confidence	0	1	2	3
52.	A change in job status	0	1	2	3
53.	Having an unexplained ache, pain, or body change	0	1	2	3
54.	_____	0	1	2	3
55.	_____	0	1	2	3
56.	_____	0	1	2	3

The next time you find yourself in a stressful situation, ask yourself this question, "Is this situation worth doing damage to my body by being stressed or upset?" Your answer may be "Yes," so go ahead, but if you continue to catch yourself, you may find that it is not worth the risk and so you save yourself the needless wear and tear of stress. Or you can ask yourself, "What is the worst thing that can happen from this?" You may realize that nothing terrible will happen! I've found that 90 percent of the things I've worried about have never materialized. How about you? Think back over your worries!

Our beliefs may exert a powerful influence over our feelings, behavior, and physical well-being. An example of how powerful our thoughts can be is illustrated by the effect of "placebos." A placebo is an inactive substance (sugar pill) or treatment that is often used to compare the effects of real drugs or treatments. Placebos have been found to be extremely potent. For many years their force was thought to be imaginary, but recently it has been demonstrated that biochemical changes, such as an increase in natural pain killing substances called endorphins do take place when people think they are receiving a pill for pain. Herbert Benson in his 1996 book, *Timeless Healing,* notes that nearly a third of experimental subjects respond to placebo and some subjects even get negative side effects from "nocebos."

Another example of the power of our thoughts and attitudes is when we are under hypnosis. There is evidence that people are able to cure warts or stop allergic reactions with hypnotic suggestions. Biofeedback also allows us greater control over physical processes. Biofeedback is the method of sending information back to the person about physical processes, which the person is not normally aware of, such as hand temperature. We don't generally notice our hand temperature unless we are very uncomfortable. Commonly when we are stressed, our hands become cooler because the blood flows away from them toward the brain and the large muscles of the body. Through biofeedback techniques, it has been found that people can acquire the ability to alter their body temperature, relax, and even avoid migraine headaches. There is more on biofeedback in the section on Relaxation Strategies.

Try this exercise and feel your hand temperature increase. Again, get comfortable and close your eyes after reading this paragraph. Imagine that you are involved with an experience from your past where your hands were very warm (for example, you are at Lake Michigan and your hands are in the sun, or you are holding your hands in front of a toasty fireplace). Just hold that image in your mind for a few moments. If you do not feel or see images very well, just repeat to yourself several times, "My hands are warm, heavy, and comfortable." With practice, you can gain control and warm your hands at will. Take a few moments and try it before finishing this section.

Much recent research has been directed toward the relationship of personality and the disease process. The clearest correlation has been in the area of "Type A" or coronary-prone behavior pattern. All the mechanisms by which the Type A pattern influences physiology have not been fully understood. It is important to note, nonetheless, that the relationship of Type A behavior pattern and the increased incidence of heart disease is strongly recognized and has been included as a risk factor for heart problems.

Two cardiologists, Meyer Friedman and Ray Rosenman, first publicized Type A behavior in the 1974 book, *Type A Behavior and Your Heart*. They define the Type A pattern as, *"an action-emotions complex that can be observed in any person who is aggressively involved in a chronic, incessant struggle to achieve more and more in less and less time."* The Type A person is characterized by an intensely competitive disposition, as well as suffering from "hurry sickness." Friedman and Rosenman are quick to point out that the Type A pattern is not psychotic or neurotically obsessed thinking as much as it is a socially acceptable form of conflict that has often been rewarded in our society. But what appears to cause substantial gains in this dog-eat-dog world apparently has substantial deleterious effects on our bodies. Most recent research tends to show that the "toxic" element of Type A is more the explosive anger and hostility than the competitiveness or time pressure.

Since such a strong case has been made for the coronary prone behavior pattern, more attempts have been made to link sets of personality traits or characteristics to other health care problems, such as cancer. Is there a cancer personality? Some researchers like Lawrence LeShan, who has been studying psychological factors and cancer since the 1950s, say there are certain tendencies exhibited significantly more in people who have developed cancer. LeShan has studied hundreds of cancer patients and found that they are much more likely to have experienced the loss of a major relationship, difficulty expressing hostile feelings, marked self-dislike, and tension over the relationship with one or both parents. They have likely experienced the loss of life's meaning or feel trapped. His books, *You Can Fight for Your Life* (1977) and *Cancer as a Turning Point* (1989) describe how cancer patients can use their disease as an opportunity to reclaim their lives. Lydia Temoshok and Henry Dreher in their 1992 book, *The Type C Connection: The Behavioral Links to Cancer and Your Health* have further described the Type C Behavior Pattern which focuses on repression and suppression of anger and "negative emotions" which produce much internal tension and pressure.

Researchers have also noted a sense of pervasive depression, or more precisely hopelessness, in people who later develop cancer. It is as though they have lost the true meaning of life or feel trapped in their life situation. Some researchers are relating long-term or major life stresses to cancer. Stress itself does not cause cancer, but as mentioned previously it may compromise the body's natural immune system.

While research has tried to link certain personality traits to specific conditions like heart disease or cancer, more evidence supports that certain "stressed" personalities are more susceptible to disease in general. Managing stress and having certain attitudes about life seem to make certain people more "hardy" or resistant to stress and disease. Personality factors may not play a role in every person's disease or health, but it may turn out that a certain subset of people may benefit by making psychological changes.

COMMUNICATING WITH YOUR MEDICAL TEAM – STRESSFUL OR REASSURING?

Being pro-active about handling stress can be especially critical as you work to communicate clearly and effectively with your medical team. You have entered a foreign and hostile land when diagnosed with cancer. "It's all Greek to me" is a phrase that fits pretty accurately. Most people feel medical jargon is another language that they don't understand and are often too scared to learn.

For instance, the word, *carcinoma*, comes from a Greek word meaning crab. That's because Hippocrates thought that breast tumors looked like crab claws piercing flesh. We get the word, *oncology*, from another Greek word, which means the study of tumors or masses.

While it may be intellectually stimulating to learn the origins of these words, medical terminology can be quite intimidating. It is vital to learn to speak "medicine" to some extent so you can communicate more effectively with your doctor and medical team. It is not within the scope of this book to be an introduction to medical terminology. There are many valuable sources of information that can help you with that, such as *Everyone's Guide to Cancer Treatment* by Dollinger, Tempero & Mulvihill detailed in the Reference section.

Communication helps you make better treatment decisions. People feel more in control when they participate in their medical care and it then helps reduce stress and anxiety. Research shows that people may have fewer problems or complications with their treatments when they are involved in making the decisions.

Some people still belong to the "old school" of desiring the doctor to make the decisions. That choice needs to be respected as well. There is nothing wrong with that path. What is most important is feeling as comfortable as possible with your doctor, medical team, and treatment decisions. As much as possible the patient's philosophy (active or passive) needs to mesh with the doctor's approach (authority or partner).

The first question is what kind of information do I need or want? How much information is enough? What is the best form of information for me to get: audio, visual, or written? Where can I find accurate information that I can trust? Some people are so overwhelmed; they don't even know where to begin. The best place to start is to ask questions of your medical team. If you are afraid you might not remember your questions, write them down. For many people a doctor's appointment is very emotional and when the doctor walks in, the patient's mind goes blank. Always take someone with you for important appointments or any appointments to help you with questions and to serve as a reality check. Take a notebook or even a tape recorder, if the doctor and you are comfortable with that. Ask for copies of your lab work or important reports. I suggest patients keep their own file so they can refer to records, as they need.

Other suggestions to help improve communication with your medical team:

- Be clear in advance about what you need to ask and tell your doctor.

- Set some priorities with your concerns so you get the most important issues covered first.

- Clarify expectations: Don't assume the doctor will ask all the right questions. You need to honestly volunteer information and ask questions that will help them provide the best care for you.

- Bring in information that you want to ask about and if you have had a troublesome symptom, keep track of it and bring those records in to your next appointment.

- Bring in your medications or keep an updated list with you, so that you and each of your doctors knows the latest medication schedule.

- Stand up for yourself appropriately and stick to the issue. Be open, honest, clear, respectful, responsible for compliance to treatment, listen and learn. Expect and request the same in return.

- Realize that passive behavior is a block to communication. Withholding information is also a block, especially if you are using complementary or alternative approaches. Many emotions can get in the way also, if they are not expressed properly. It may be difficult, but communicating can increase satisfaction with your patient-doctor relationship.

Psychoneuroimmunology

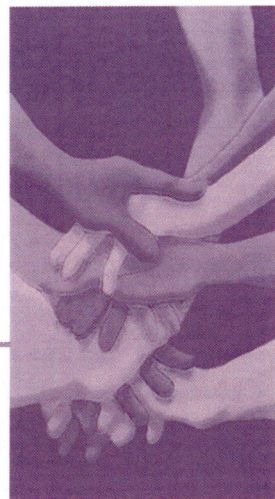

(PNI)

A FEW WORDS ABOUT PSYCHONEUROIMMUNOLOGY (PNI)

Not everyone who is exposed to a germ catches a disease. Our ability to fight off an illness relates to a number of diverse factors. Strengthening our **host resistance** helps make it more difficult to get sick in the first place and speeds our recovery. Lifestyle factors such as diet, exercise, tobacco and drug use, as well as spiritual connection, genetics, and environmental exposure make up the whole person. These factors interact and lead to any state of health or illness as in the diagram below.

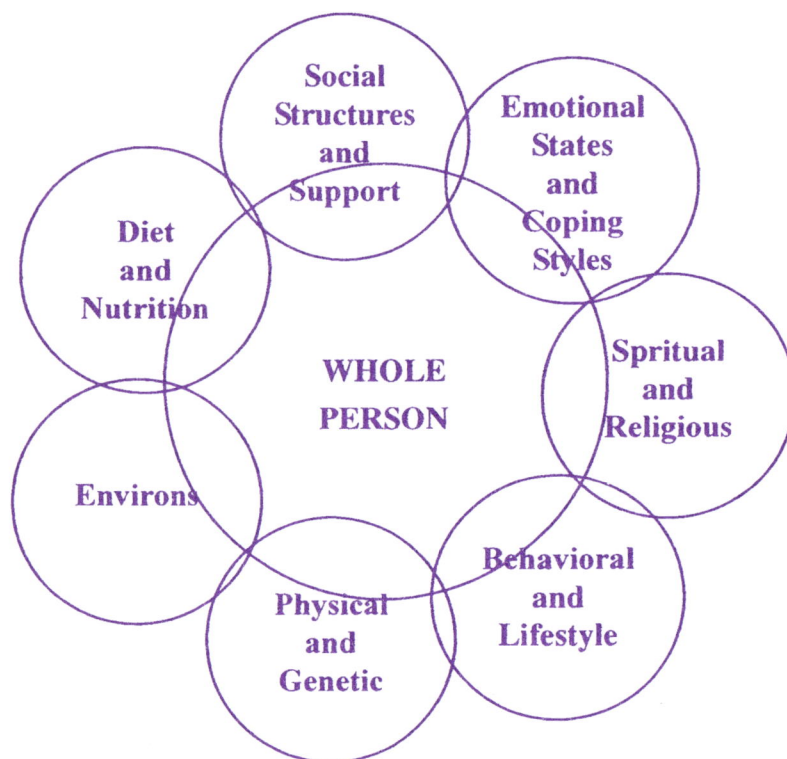

Social Structures and Support

Emotional States and Coping Styles

Diet and Nutrition

WHOLE PERSON

Spritual and Religious

Environs

Behavioral and Lifestyle

Physical and Genetic

Cancer is a complex health problem that illustrates this. We may be born with a genetic predisposition due to faulty genes (**oncogenes** and **tumor suppression genes**), but that may not be enough to cause cancer. The more exposure to cancer causing substances (**carcinogens**) such as radon, chemicals, tobacco smoke, etc. the more damage is done to the DNA of the cells changing them from normal to healthy cells. Stress and personality factors may be involved not directly in damaging the DNA of cells, but of decreasing the number of immune cells or their ability to discover cancer and get rid of it.

Psychoneuroimmunology **(PNI)** is a new area of study that examines the interaction of our nervous system **(NS)**, endocrine system **(ES)**, and immune system **(IS)**. In other words, how psychological (mental) processes affect our bodies' defense systems, which are physical.

For a number of years, the immune system was viewed as autonomous, that is, acting on its own. In the mid-70s, researcher Robert Ader ignited interest in how the immune system could be affected by conditioning. Ader was studying taste aversion in rats. He was pairing a sweet solution with a drug that produced nausea to see if the rats would avoid the sweet liquid. He happened to notice serendipitously (by accident or coincidentally) that many rats were dying. It turned out that the nausea-producing drug also lowered the immune system. The immune system continued to get lower even without the drug, whenever the rats tasted the sweet solution. In other words, the rats' immune systems became conditioned to decrease. This jargon meant that the immune system could be affected by psychological processes. Ader wondered whether the immune system could be increased through a conditioning process and also wondered whether humans could do it. The answer appears to be yes to both questions and research in this area has expanded dramatically in the last twenty-five years. **PNI** has developed as a legitimate area of study during those decades.

We have since gained evidence that the **IS** directly communicates with both the **NS** and **ES**, and that information is exchanged bi-directionally. For example, there are receptors for neurotransmitters (and other chemical messengers) on the surfaces of immune cells. Neurotransmitters are substances that allow the nerves to talk to one another. These receptors pick up neurotransmitters so the immune cells are able to gain information from our nervous system and our brain. This means that mind-body connections are not just theoretical but an anatomical reality as is represented in the figure below.

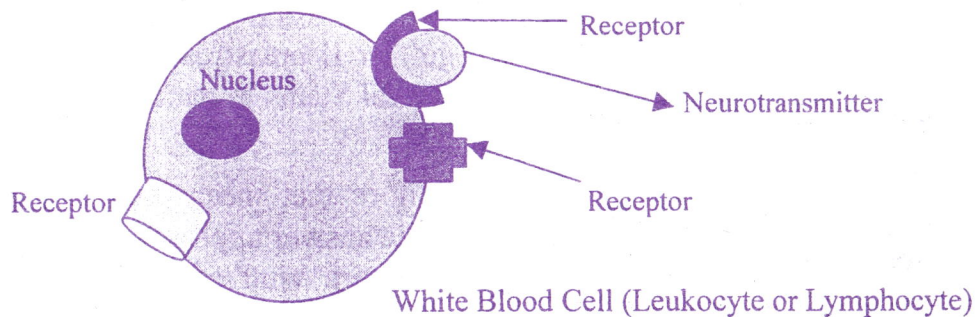

White Blood Cell (Leukocyte or Lymphocyte)

Likewise, our autonomic nervous system links directly to lymph tissue, which is an important filtration mechanism to trap foreign substances and rid the body of them. There is also a body of evidence that relates how the immune system is affected by psychological processes. A direct case in point is that our bodies' defenses are both qualitatively and quantitatively weakened by stress, especially when we are not coping well or using relaxation techniques like those you will learn from this manual.

Many different kinds of stress have been studied, such as:

❶   Acute laboratory stress, like mental math problems
❷   Major life stress or ongoing stress, like marital conflict
❸   Bereavement, especially death of a spouse
❹   Mood states, such as depression or anxiety
❺   Natural disasters, such as hurricanes and earthquakes
❻   Hassles, i.e. those minor daily irritants that tend to add up

Recently, Janice Kiecolt-Glaser and her colleagues ("Psychological Influences on Surgical Recovery Perspectives From Psychoneuroimmunology," *American Psychologist*, 1998, 53, 11, 1209-1218) noted the effects of stress on wound healing. Her study showed that caregivers for Alzheimer's patients took nine days or 24 percent longer than people who were not caregivers to heal a small, standardized wound. One of her fellow researchers, Phillip Marucha, mentioned in the same article, noted that small oral wounds in the mouths of dental students took 40 percent longer to heal during exam week than during vacation. These authors concluded that stress might affect several biochemical processes that slow healing.

Another interesting study was done by Dana H. Bovbjerg and other researchers at Memorial Sloan-Kettering Cancer Center ("Anticipatory Immune Suppression and Nausea in Women Receiving Cyclic Chemotherapy for Ovarian Cancer," *Journal of Consulting and Clinical Psychology*, 1990, 58, 2, 153-157). They found that white blood cell response was lower than several days before, when 20 ovarian cancer patients entered the hospital for their treatments. They interpreted that to mean the immune system had become conditioned downward after having several chemotherapy treatments that decreased the immune system. Tjemsland and colleagues found that intrusive anxiety and depression were related to decreases in B and T lymphocytes in studying 106 Stage I or II breast cancer patients ("Properative Psychological Variables Predict Immunological Status in Patients with Operable Breast Cancer," *Psycho-Oncology*, 1997, 6, 311-320).

Just because stress impairs the **IS,** does that mean increasing our ability to manage stress improves **IS** functioning? The answer appears to be **YES**. A variety of interventions may help increase one or more vital components in the immune system's defenses. These techniques such as the following:

❶   Muscle Relaxation
❷   Guided Imagery (Visualization)
❸   Hypnosis and Self-Hypnosis
❹   Cognitive-Restructuring (Attitude Adjustment)
❺   Disclosure (Journaling, Psychotherapy, Constructive Venting)
❻   Support Group Attendance or Spiritual Connection
❼   Exercise, Yoga, Massage, Laughter

An unpublished dissertation by Edwards and his colleagues, presented in 1990, showed that breast cancer patients who were taught relaxation-visualization to imagine increasing their natural killer cells did average a 12 percent increase in the activity of the immune system's killer cells when practicing that exercise daily. Other researchers like Barry Gruber and his associates ("Immune System and Psychological Changes in Metastatic Cancer Patients Using Relaxation and Guided Imagery: A Pilot Study," *Scandinavian Journal of Behavior Therapy*, 1988, 17, 25-46 and "Immunological Responses of Breast Cancer Patients to Behavioral Interventions," *Biofeedback and Self-Regulation*, 1993, 18, 1-21) have shown similar results with a wide variety of immune measures in patients with different types of cancer.

Another recent study by Lekander and other researchers ("Immune Effects of Relaxation During Chemotherapy for Ovarian Cancer," *Psychotherapy and Psychosomatics*, 1997, 66, 4, 185-191) showed a small group of ovarian cancer patients were able to increase their lymphocyte counts with relaxation training even while on active treatment. Ongoing research by Barbara Andersen and other researchers at Ohio State University ("Stress and Immune Responses After Surgical Treatment for Regional Breast Cancer," *Journal of the National Cancer Institute*, 1998, 90, 30-36) has shown the detrimental effects of stress on Natural Killer (NK) cells as well as examining the positive effects of a psychological and behavioral program with Stage II and III breast cancer patients.

Another study by Shrock, *et al.* in *Alternative Therapies*, 1999, 5, 3, 49-55, "Effects of a Psychosocial Intervention on Survival Among Patients with Stage I Breast and Prostate Cancer: A Matched Case-Control Study" found increased survival time in both breast and prostate cancer patients who had been in six, two-hour group sessions. Numerous other studies have found beneficial effects on sleep, anxiety, depression, pain, general psychological adaptation and even lowered medical costs. A study done in Canada (Simpson, Carlson, & Trew, 2001) found that a cognitive-behavioral series of six classes found a reduction in depression and other psychiatric symptoms as well as improvement in quality of life and lower medical expenses after they completed treatment. Michael Antoni and his group at the University of Miami, Coral Gables, Florida has also found a cognitive-behavioral group approach helpful to breast cancer patients both physically and emotionally.

Although I have been unable to collect immune measures, I have run several sets of classes to help teach breast cancer patients stress management strategies. Using this manual as a workbook, we met for six, two-hour sessions and the participants found it helped lower anxiety and depression, as well as increased a feeling of support and hope. There will be more on the role of support in a later chapter.

Our physical health may also benefit from greater constructive expression of feelings. Holding feelings in may add greater stress on the body, especially the immune system. Expressing your self through discussions with family and friends, attending a support group, individual therapy, art, and writing may be constructive ways to process and resolve conflicts and feelings to release tension and improve physical health. The topic of creative expression and journaling is discussed in another chapter in greater depth. One study that appeared in the October 15, 2002 *Journal of Clinical Oncology* found that women who recorded their "deepest thoughts and feelings" reported fewer physical symptoms, including coughing and sore throats. They also had fewer unscheduled appointments for cancer–related illness.

The area of spiritual expression and immune function has also been explored. A study that examined 112 women with metastatic breast cancer found that helper and cytotoxic t-cell counts were greater among women reporting greater spiritual involvement as measured by frequency of attendance and reported importance to their lives. Another study at Duke University called "The MANTRA Project" found that heart patients who received prayer have 50 to 100 percent fewer side effects. First names only were sent via email to Buddhists, Hindus, Jews, Protestants, and Catholic nuns who regularly prayed for the patients in the experimental group.

Two other studies of interest were reported by British researcher, Professor Leslie Walker, and his colleagues at the British Psychosocial Oncology Society Annual Conference in 1996. The first study examined 96 women with large (>4 cm) or locally advanced tumors and their response to chemotherapy. Psychological responses were documented using a number of standardized tests. Women who had high anxiety and high external locus of control with larger tumors had a poorer response to treatment. The same kind of relationship occurred with women who had larger tumors and higher depression scores.

In their other study, the researchers found that relaxation and guided imagery had a positive impact on several immune measures. Eighty women used the techniques while undergoing chemotherapy and radiation treatments. There was a practice effect noted. In other words, the more one practiced, the more benefit might be experienced.

We cannot measure to what extent psychological processes can affect our physical functioning in the onset of cancer or in recovering from it, but it is a very exciting area of exploration. Another important factor to keep in mind is that different cancers may have different effects on our immune system and can fool our immune system in several ways because it arises from our own cells. It is gratifying for researchers, but more importantly, it offers hopeful evidence to cancer survivors and their families that we can have a positive and sometimes profound affect on our own recovery and well-being. A possible model of how cancer develops and how it might regress with psychological interventions is shown on the following pages.

Later in this chapter, you will find a list of the major components of the immune system. Following are tools designed to help you become more aware of what factors may adversely or positively affect the human immune system. Keep in mind that certain cells of the immune system are more involved in cancer than others. A last important concept is that our immune system works in a very delicate balance. If it is decreased it may make us more susceptible to disease and if it is too active then it may attack our own body as it does in autoimmune health problems such as Multiple Sclerosis, Rheumatoid Arthritis, Lupus, and some forms of Diabetes. Do not just bolster your immune system indiscriminately. Work with professionals and your medical team to use these suggestions to help improve your body's defenses.

THE FOUR WAYS PNI CONTRIBUTES TO HOLISTIC CANCER CARE

1. Helps Patients Cope and Restore Control

Patients adjust to diagnosis more easily
Patients have less anxiety and depression
Patients may reduce pain levels and improve sleep

2. Helps Increase Treatment Compliance and Symptom Relief

Patients are more likely to receive therapeutic treatment benefits
Patients are more likely to reduce pre- and post-treatment effects

3. Helps Motivate Patients to Change Lifestyle Behaviors in a Healthy Direction

Patients are more likely to improve
Diet, exercise, and cessation of smoking
Patients are more likely to find benefit or experience post-traumatic growth

4. May Help Improve Treatment Outcome and Affect Quality of Life or Survival in a Positive Direction

May reduce medical expenses
Patients may experience improvement in immune measures
Patients are likely to be more active and achieve more realistic goals
May improve treatment outcome

MINDBODY MODEL OF
CANCER DEVELOPMENT

This model (first proposed by Jeanne Achterberg and G. Frank Lawlis in 1978) is not intended as a comprehensive model of how cancer develops. Genetic, physical, dietary, and environmental factors are not included in this diagram. However, it shows that psychologically stressful events can affect several brain areas, which then adversely deregulate the immune and hormonal systems, which may then lead to cancer growth.

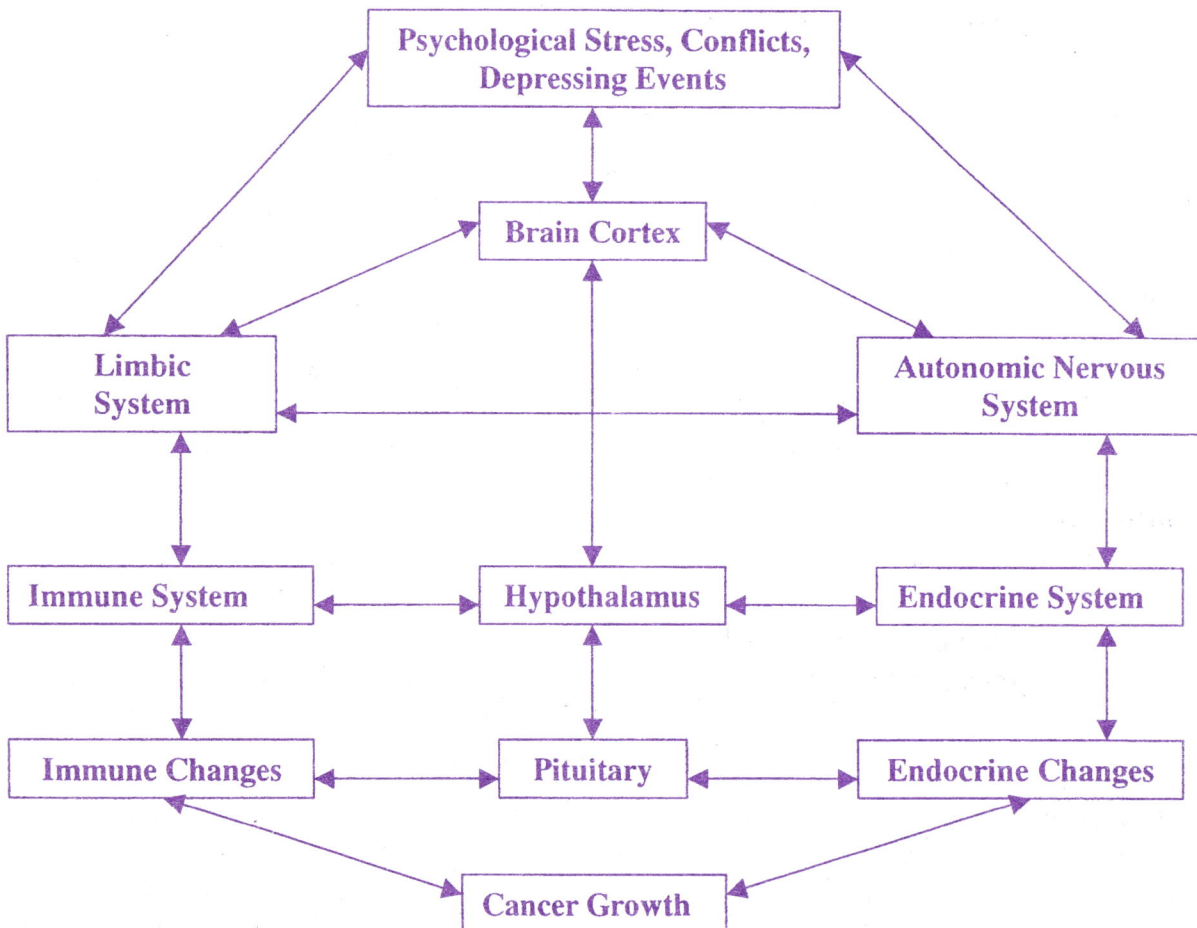

```
        Psychological Stress, Conflicts,
             Depressing Events
                    │
                Brain Cortex
         ┌──────────┼──────────┐
      Limbic                Autonomic Nervous
      System                    System
         │                        │
   Immune System ── Hypothalamus ── Endocrine System
         │              │              │
  Immune Changes ── Pituitary ── Endocrine Changes
              └──── Cancer Growth ────┘
```

Note: All communication is bi-directional. Communication among the three systems is achieved by **neurotransmitters** of the Nervous System, such as acetylcholine and norepinephrine; **hormones** of the Endocrine System such as adrenocorticotrophin (ACTH) and corticotrophin releasing hormone; and Immune System **cytokines** such as interleukin-1 (IL-1), interferon, and tumor necrosis factor (TNF).

MINDBODY MODEL OF
CANCER REGRESSION

If certain attitudes, mood states, beliefs, and difficulty coping with stress can contribute to cancer occurrence, then reversing or altering those might aid treatment outcome and cause cancer to regress. It is difficult to predict how much psychosocial and spiritual interventions may positively impact an individual's recovery. For some, it may be a significant factor. **Again, the diagram below is not intended to be comprehensive, but reflects the role psychoneuroimmunology can play in cancer regression.**

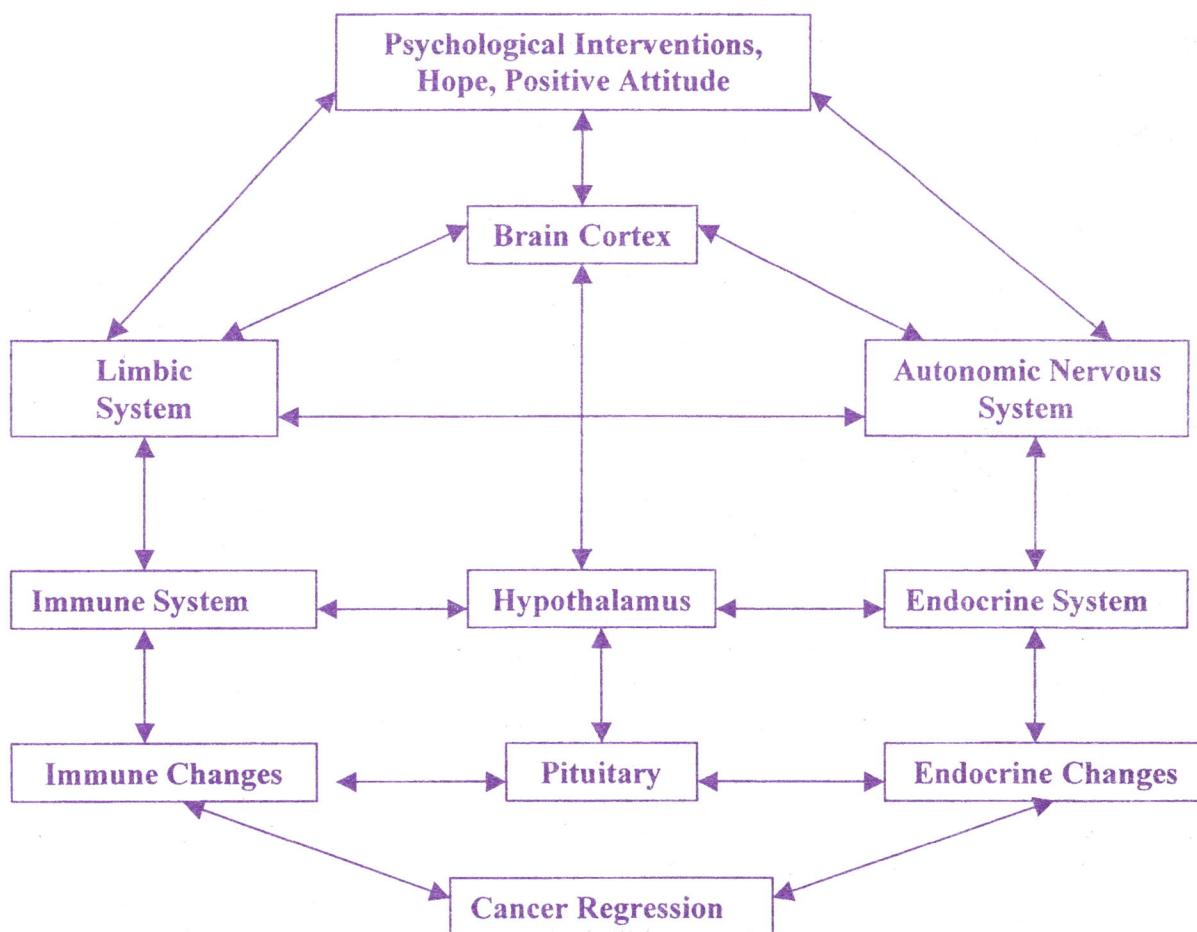

```
            ┌─────────────────────────────┐
            │ Psychological Interventions, │
            │    Hope, Positive Attitude   │
            └─────────────────────────────┘
                        │
                ┌───────────────┐
                │  Brain Cortex │
                └───────────────┘

┌──────────────┐                    ┌──────────────────┐
│    Limbic    │◄──────────────────►│ Autonomic Nervous│
│    System    │                    │      System      │
└──────────────┘                    └──────────────────┘

┌──────────────┐   ┌──────────────┐   ┌──────────────────┐
│Immune System │◄─►│ Hypothalamus │◄─►│ Endocrine System │
└──────────────┘   └──────────────┘   └──────────────────┘

┌──────────────┐   ┌──────────────┐   ┌──────────────────┐
│Immune Changes│◄─►│  Pituitary   │◄─►│ Endocrine Changes│
└──────────────┘   └──────────────┘   └──────────────────┘

            ┌─────────────────────┐
            │  Cancer Regression  │
            └─────────────────────┘
```

DEFINITIONS OF MAJOR IMMUNE COMPONENTS

Pluripotent Stem Cell – produces all different blood cell lines

Granulocytes – nonspecific defending cells that attack any type of invader

> **Neutrophil** – guard against infection
> **Eosinophil** – involved in allergic reactions, kill parasites
> **Basophil** – involved in allergic reactions, release histamine

Monocytes and Macrophages – engulf invaders

T-Cells (Thymus Derived)

> **Helper T-Cells** – signal B-cells to secrete antibodies
> **Suppressor T-Cells** – counteract Helper T-cells
> **Memory T-Cells** – remember a particular invader
> **Killer T-Cells** – kill specific infected or cancer cells

B-Cells (Bone Marrow Derived)

> **Plasma Cells** – secrete specific antibodies
> **Memory Cells** – remember a specific antigen or invader

Natural Killer (NK) Cells – kill nonspecific abnormal cells, such as cancer cells

Cytokines – chemical messengers that allow blood cells to communicate via signals to increase or decrease production or growth

Dendritic Cells – help stimulate other immune cells being examined as possible vaccine

IMMUNE SYSTEM "DOWNERS"

Morry Edwards, Ph.D.

This assessment reviews all the factors that have been known to decrease one or more immune system components or activities. This is not a scientific instrument nor is it intended to diagnose. Instead, it is a tool to help you become more aware of how you might be lowering your body's defenses. You then may make lifestyle changes to improve your immune system. Please mark the appropriate column and total your score. Not Present = 0, Seldom = -1, On Occasion = -2, Frequent = -3. A higher negative score means a more stressed immune system.

FACTOR	Not Present	Seldom	On Occasion	Frequent	Score
MAJOR STRESS					
Work	0	-1	-2	-3	
Marriage	0	-1	-2	-3	
Immediate Family	0	-1	-2	-3	
Extended Family	0	-1	-2	-3	
Loss	0	-1	-2	-3	
Injury	0	-1	-2	-3	
Legal	0	-1	-2	-3	
OTHER (Specify Below)					
A)	0	-1	-2	-3	
B)	0	-1	-2	-3	
C)	0	-1	-2	-3	
HASSLES (Specify Below)					
A)	0	-1	-2	-3	
B)	0	-1	-2	-3	
C)	0	-1	-2	-3	
Depression	0	-1	-2	-3	
Anxiety	0	-1	-2	-3	
Lack of Coping Skills	0	-1	-2	-3	
SUBTOTAL 1:					

FACTOR	Not Present	Seldom	On Occasion	Frequent	Score
PERSONALITY					
Pessimistic	0	-1	-2	-3	
Low Self-Esteem	0	-1	-2	-3	
Avoidant	0	-1	-2	-3	
Dependent	0	-1	-2	-3	
Worrisome	0	-1	-2	-3	
Inhibited/Suppressed/Repressed	0	-1	-2	-3	
Lack of Social Support	0	-1	-2	-3	
Lack of Spiritual Support	0	-1	-2	-3	
Lack of Cultural Identity	0	-1	-2	-3	
Feeling Unloved	0	-1	-2	-3	
Problems Without Solutions or Therapy	0	-1	-2	-3	
POOR GENES					
Early Death or Illness Immediate Family (2x Each)	0	-1	-2	-3	
Early Death or Illness Extended Family (2x Each)	0	-1	-2	-3	
PHYSICAL SYMPTOM					
Headache	0	-1	-2	-3	
Chest Pain	0	-1	-2	-3	
GI Distress	0	-1	-2	-3	
Other:	0	-1	-2	-3	
Other:	0	-1	-2	-3	
SUBTOTAL 2:					

FACTOR	Not Present	Seldom	On Occasion	Frequent	Score
Infections	0	-1	-2	-3	
Allergies	0	-1	-2	-3	
Age	0 (16-40s)	-1 (40-55)	-2 (55-65)	-3 (>65)	
Smoking	0	-1	-2	-3	
POOR SLEEP					
Trouble Falling Asleep	0	-1	-2	-3	
Frequent Awakenings	0	-1	-2	-3	
Restless Sleep/Bad Dreams	0	-1	-2	-3	
Early Awakenings	0	-1	-2	-3	
< 8 Hours	0	-1	-2	-3	
POOR DIET					
High Fat	0	-1	-2	-3	
High Sugar	0	-1	-2	-3	
Unbalanced	0	-1	-2	-3	
High Junk Food	0	-1	-2	-3	
Low Fruits and Veggies	0	-1	-2	-3	
Obesity	0	-1	-2	-3	
Inactive or Excess Activity	0	-1	-2	-3	
DRUGS					
Alcohol	0	-1	-2	-3	
Marijuana	0	-1	-2	-3	
Cocaine	0	-1	-2	-3	
Caffeine	0	-1	-2	-3	
Uppers/Diet Pills	0	-1	-2	-3	
Steroids	0	-1	-2	-3	
Antibiotics	0	-1	-2	-3	
Pain Killers	0	-1	-2	-3	
Muscle Relaxants	0	-1	-2	-3	
SUBTOTAL 3:					

FACTOR	Not Present	Seldom	On Occasion	Frequent	Score
Decongestants	0	-1	-2	-3	
Other:	0	-1	-2	-3	
Other:	0 (16-40s)	-1 (40-55)	-2 (55-65)	-3 (>65)	
ENVIRONMENTAL					
Hazardous Chemicals	0	-1	-2	-3	
Heavy Metals	0	-1	-2	-3	
Pesticides	0	-1	-2	-3	
Radiation	0	-1	-2	-3	
Ill People	0	-1	-2	-3	
Other:	0	-1	-2	-3	
Other:	0	-1	-2	-3	
Other:	0	-1	-2	-3	
SUBTOTAL 4:					
SUBTOTAL 3:					
SUBTOTAL 2:					
SUBTOTAL 1:					
GRAND TOTAL:					

IMMUNE SYSTEM "UPPERS"

Morry Edwards, Ph.D.

This table lists all the factors that can possibly improve one or more components or activities of the immune system. This is not intended as prescriptive. You are encouraged to explore the research and incorporate changes that help you feel better, give you more energy, or help prevent illness. Obviously, you should avoid as many **Immune System "Downers"** as possible.

Good Genes → Most family members die at a late age (85 or older) of so-called "natural causes." Little or no family history of cancer or autoimmune diseases (e.g. Lupus, MS)

Age → Our Immune System functions best from our mid-teens until our early 40s. It is not fully developed until after puberty starts and begins to decline as we enter our 60s.

Few Frequent Symptoms → Health problems such as Headache, GI Distress

Few Repeated Infections → Not prone to colds, flu, bronchitis, etc.

Few Allergies or Sensitivities → Especially common substances

Balanced Hormones → Treat any hormone deficiencies or imbalances

Exercise → This means moderate to vigorous exercise of twenty to thirty minutes at least three or four times a week

Regular Relaxation → Minimum of one-half to one hour a day engaged in activity that you find calming or quieting (e.g. meditation, movement, reading, gardening, etc.)

Acquire Coping Skills → Improve self-confidence and assertiveness

Improve Social Support → Improve the quality of current relationships and/or increase the number of positive relationships you have

Develop Awareness → Slow down and become more fully conscious of the world around you (Mindfulness)

Regular Practice of Quiet → Develop a routine of meditation, relaxation, or prayer

Regular Practice of Joy → Develop routine involvement in hobbies, crafts, or areas of interest, which promote enjoyment and affirmation. Expand your horizons!

Increase Your Altruism → Develop ways you can give to others without feeling resentful.

Confront Your Fears → Stop tying up energy being afraid and tense. Begin to view things as a challenge and take at least small steps toward conquering that.

Resolve or Neutralize Sources of Anger → Unresolved or unexpressed anger burns energy. Learn to let go in some constructive way.

Improve Balance In Your Diet → Increase variety and make sure proteins, carbohydrates and beneficial fats are in proper proportion. Maintain an acid/alkaline balance.

Improve Gastrointestinal Health → Cleanse and detox your system from time to time. Maintain even flow.

Increase Fiber in Diet → Make sure you add both soluble and insoluble fiber and reduce refined food consumption.

Reduce/Avoid Sugar In Diet → Limit or reduce refined carbohydrates.

Reduce Unhealthy Fats → Limit your diet to 20-25 percent healthy fats, such as Monounsaturated and Omega-3 fatty acids. Avoid Saturated Fats, Trans-Fatty Acids, and Hydrogenated Fats.

Increase Vegetables/ Fruits → These foods contain phytochemicals especially antioxidants, which increase immunity.

Avoid Problematic Foods → These are foods that create unpleasant symptoms or sensations after ingestion. Foods that you are allergic or sensitive to.

Consider a Supplement → If you miss meals or have an unbalanced diet, a high-potency, balanced, hypoallergenic, comprehensive multivitamin/mineral supplement may be beneficial.

The Following Table Shows Some Beneficial Immune Nutrients

Vitamin A / Carotenoids	Maitake, Shiitake, or Reishi Mushrooms
Vitamin B1, B2, B3, B 5, B6, B12	Green tea
Folic Acid, Biotin, Choline	Echinacea
Inositol, PABA	Astragalus
Vitamin C	Berberis
Bioflavonoids	Ganoderma (reishi mushroom)
Vitamin D	Gentiana
Vitamin E (mixed tocopherols)	Ginseng (Siberian,Korean,American)
Calcium Ho	Shou Wu
Copper	Licorice (not candy)
Magnesium	Ligustrum
Selenium	Rehmannia
Zinc	Schisandra
CoQ10	White Atractylodes
Garlic	Modified Citrus Pectin
Glutathione	Flaxseed Oil
Lipoic Acid	Soy Isoflavones
L-carnitine	Molybdenum
Quercetin	Beta 1,3-Glucan
Germanium	MGN-3

Four very helpful resources in this area are the following:

Kenneth Bock, M.D. and Nellie Sabin, *The Road to Immunity: How to Survive and Thrive in a Toxic World*, New York: Pocket Books, 1997.

Donald R. Yance, Jr., C.N., M.H., A.H.G. and Arlene Valentine, *Herbal Medicine, Healing & Cancer A Comprehensive Program for Prevention and Treatment*, Lincolnwood (Chicago), IL: Keats Publishing, 1999.

Kedar N. Prasad, Ph.D., *Vitamins in Cancer Prevention & Treatment: A Practical Guide*, Rochester, VT: Healing Arts Press, 1994.

Patrick Quillin, Ph.D., R.D., C.N.S. and Noreen Quillin, *Beating Cancer with Nutrition*, Carlsbad, CA: Nutrition Times Press, Inc., 2001.

INFORMED INCENT (IVE)

You have already been informed of the possible aversive effects of your treatment. Let it be known that complying fully with your treatment may allow you to enjoy the following positive outcomes:

1. To laugh out loud for no apparent reason.

2. To appreciate the absolute silliest or simplest of things.

3. The potential to see your children or grandchildren grow up.

4. The chance to create that masterpiece you've always wanted or at least produce something artistic that pleases you.

5. The possibility that you will take that journey you've always wanted.

6. The option to connect or reconnect with people who have meant something to you.

7. The likelihood you will examine all areas to make you even healthier.

8. A meaningful chance to explore priorities and spiritual avenues to connect with the Source of the Universe.

9. The powerful flow of energy pulsing through your body.

10. The sight of the sun on the water.

11. The gentle feel of the breeze against your cheek.

12. The taste of chocolate, ice cream, or pizza.

And room for a whole host of other life-affirming activities that you can list here as goals you wish to reach.

_____

_____

_____

_____

_____

Signature _____

Date: _____

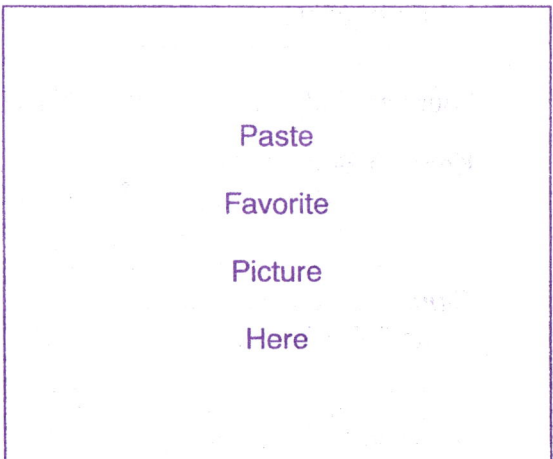

Paste

Favorite

Picture

Here

Relaxation Strategies

RELAXATION AND THE RELAXATION RESPONSE

People react to different situations as stressful and people exhibit stress in a variety of ways. The same is true of relaxation. Different techniques work uniquely for individual people. It's like the Gary Larson cartoon that shows one bear talking to another about how relaxed he gets coming into the city and standing on a busy street corner. While a busy street corner is unlikely to be relaxing, the point is that there are lots of activities, which could potentially be relaxing. Even though people may benefit from inner focused relaxation strategies like deep muscle relaxation or meditation, not everyone reacts positively to those or will take time to do them.

Whether you have cancer or not, our complex and fast paced world is not going to slow down unless we take the time to slow it down. We need to schedule some relaxation or down time regularly and regularly means **daily**. This "conscious slowing" will allow our bodies to recharge and restore balance.

The first step is to make sure you have some relaxing activities. You may find meditation, journaling, reading novels, exercising, gardening, arts, crafts, watching your fish tank, or just plain "vegging out" calms you right down. If you do not have any activity that brings joy or calmness, then the first step is to find something.

The second step is to allow your self a minimum of one-half to one hour a day to enjoy these activities. If you feel guilty because you are not being "productive," remember you are doing three important things when you relax:

❶   You are restoring balance in your body by shifting into your parasympathetic nervous system (PNS).

❷   You are recharging energy that is needed for your real priorities.

❸   You are improving your body's defenses.

These benefits are in addition to the positive psychological changes you can also experience, such as less irritability, greater enjoyment of life, and feeling less rushed and anxious.

In this manual we are focusing on relaxation techniques that would not be considered recreational. It is important to take time to focus inwardly so you can consciously teach your body to relax and learn what that feels like. Then, you can call back (control) that response, when you begin to feel yourself getting anxious.

At this point, it is important to make a distinction. While sleeping or taking a nap may be necessary and helpful, focused relaxation is much different in two ways. First, when you relax, you consistently slow your body down more than when you sleep.

Herbert Benson (and other researchers as well) relates in *Timeless Healing* that metabolism in people who are meditating slows 25 percent more than when those same individuals are sleeping. Sometimes sleep itself is not very relaxing, especially when there are vivid dreams, restlessness, or breathing problems. Second, and more importantly, you become conscious of how you actually feel and what relaxes you, so you can control your reactions better. I am not suggesting you forego naps as they can certainly recharge you. I am just suggesting that focused relaxation and sleep serve different purposes.

One other important point: **DO NOT TRY TO RELAX!** Again, it is like the Gary Larson cartoon where the therapist is encouraging the unsuccessful chameleon not to try too hard to change colors. The harder you try to relax, the more tense and frustrated you are likely to get. Just let it happen by "letting go." "Letting go" is something you have to experience for yourself.

Think of different relaxation techniques or strategies — muscle relaxation, visualization, meditation, self-hypnosis, etc. — as structures that make letting go easier. There are any number of relaxation tapes on the market, one of which may appeal more than others, so try several and experience what works best. You might like a soft voice to guide you through an exercise. You might prefer gentle nature sounds or slow instrumental music. Steven Halpern's *Spectrum Suite* is a good example of specially engineered music to help you achieve a meditative state. Just focusing on your breathing, gently slowing down and falling into a comfortable rhythm, may be all you need.

I have included a transcript of a relaxation tape that you can make for yourself. (If you don't care for the sound of your own voice, ask someone else to tape it.) If you don't have a tape recorder, you can memorize the general idea of systematically relaxing one part of your body at a time. If twenty to thirty minutes is too long, simply begin by quieting yourself down for two or five minutes. In practicing deep, focused relaxation, there are four main considerations:

❶ A QUIET AND COMFORTABLE ENVIRONMENT

Try to find a place that will become your **relaxation or meditation space.** Temperature, lighting, and ventilation should be conducive to relaxation. You may have a pleasing view or smell that you find comforting. Minimize distractions by unplugging the phone, keeping the cat or dog out, and posting a "RELAXATION IN PROGRESS, DO NOT DISTURB" sign. That way other family members may not bother you or not be mystified with what you are doing. Establish a routine. Although you can be flexible, it is easier for most people to learn to relax and to maximize its benefits if they develop a routine with an established time and place. You will find yourself quickly becoming conditioned and notice your body and mind relax when you enter your "space."

❷ A COMFORTABLE POSITION

Usually this involves sitting in a comfortable chair like a recliner or lying in bed. You may prefer to sit (or recline) on the floor. Wear loose clothing and position yourself in such a way that as many muscles as possible are relaxed and supported. Although you may want to use the relaxation strategy to help you sleep at night, you shouldn't fall asleep often when doing the relaxation. If you are lying down, then change to a sitting position. If you fall asleep sitting up, adjust that position. Remember you need to learn what the relaxation response feels like so you can control it, when you need to relax.

❸ A MENTAL DEVICE

The mind has been compared to a "drunken monkey." Therefore, a structure helps focus our mind and quiet it down. Especially when our thoughts are racing, we need something that will "bring us back to our center." Trying to grit your teeth and make your mind behave often makes it go that much faster. You may want to start by **anchoring** your attention. Staring at a focal point or object with your eyes open is one strategy many use to successfully "anchor" their attention. An example would be a picture on the wall or a colorful object on your desk. If your attention wanders, bring it back to that spot. If this is too distracting, then close your eyelids while gently looking up at the center of your forehead. Just following your breathing for a few minutes may noticeably slow you down. You may want to use a taped exercise or memorize an exercise that you can imagine yourself. Some people like to focus on a blank screen or follow their thoughts without becoming emotionally attached.

As we slow down and get quiet, all the "static" is tuned out and we can access our inner wisdom. Our intuitive understanding is often necessary to balance our rational mind in helping make balanced decisions. At some deep level, we know what is best for us and what we truly need. We must get quiet to really connect with that inner wisdom and hear what is healthy for us. A way to make the relaxation more transportable is to use a **cue word**. When you notice internal sensations that feel calm, then say a particular word like "RELAX," "PEACE," or "LET GO." By pairing the word and the sensations many times, they will become associated or conditioned. When you notice yourself becoming tense, all you will have to do is say the Cue Word and you will feel yourself become calmer. You can also have a non-verbal cue like touching your thumb and middle finger together. Just remember to do it often so that the conditioning occurs.

❹ A POSITIVE AND PASSIVE ATTITUDE

Your daily quiet time or relaxation practice must not just be another item on your "to do list." It is best if you genuinely look forward to this time. If you consciously recognize that it is advantageous to your health and well-being, you can enlist your relaxation time as a kind of ally in your fight against cancer. Remember, do not try to make yourself relax. Typically, it will have the opposite effect and you will become more frustrated and tense.

Various relaxation, meditation, and visualization practices such as those outlined above have been found to help cancer patients and their family members cope with a variety of problems that can arise during their disease.

Researchers like Thomas Burish and associates at Vanderbilt University ("Preparing Patients for Cancer Chemotherapy: Effect of Coping Preparation and Relaxation Interventions," *Journal of Consulting and Clinical Psychology*, 1991, 59, 518-525) and William Redd and associates ("Hypnotic Control of Anticipatory Emesis in Patients Receiving Cancer Chemotherapy," *Journal of Consulting and Clinical Psychology*, 1982, 50, 14-19) now at Memorial Sloan-Kettering have found that learning deep muscle relaxation (active or passive) improves sleep, reduces anxiety, decreases complications and side effects from chemotherapy and radiation treatments, and increases treatment compliance for full therapeutic effectiveness.

As mentioned earlier in the section on Psychoneuroimmunology (PNI), guided imagery and visualization have shown an increase in certain immune blood cells (**Natural Killer Cells**), although the effect on outcome and cancer survival remains controversial. Several studies with cancer patients (including work by this author) both on and off treatment have found immune function enhanced. Hypnosis with imagery suggestions has been especially useful as a pain control strategy.

Prayer has recently been studied for its effects on our health. In three significant studies, prayer has been found to reduce complications with cardiac patients, decrease hypertension, and benefit immune measures in AIDS patients. There is no reason to suspect that cancer patients wouldn't also benefit. Finding meaning from the experience of cancer and using it to make life-affirming changes has been associated with improved treatment outcome.

Pyschoeducational and support groups have also reduced distress and increased coping skills. Somewhat more controversial have been findings that show improvement in immune parameters and increased survival. At least four well-conducted studies have demonstrated improvement with several different cancers (breast, prostate, melanoma, and gastric).

On the following pages is an overview of particular stress management strategies. Many of these are described more fully, so you can try them and experience which work most effectively for you. You do not need to feel as though you have to master them all. See if two or three work well and practice them regularly. Think of this resource manual as an introduction, intended to expose you to a wide variety of ideas that you can experiment with and tailor to your needs.

A List of Some Useful Stress Management Strategies

I. **Deep Relaxation** (20-30 minutes, once or twice a day)

- Active – **Progressive Muscle Relaxation (PMR)**
  Alternating tension and relaxation of specific muscles

- Passive – **Autogenic Training (AT)**
  Self-generated phrases for relaxation

- **Relaxation Response (RR)** – Like **Transcendental Meditation (TM)**, but Westernized

- **Biofeedback (BF)** – Using bioelectronic equipment to monitor a physical process like muscle tension and translate that information to the person to control it better

II. **Brief Relaxation Strategies** (2-5 minutes, several times a day, when needed)

- **Feeling Pause (FP)** – Stop to assess "stress barometer" and learn to intervene before stress gets out of control

- **Body Stress Scanning (BSS)** – Same as **FP** but scan your whole body instead of "stress barometer"

- **Diaphragmatic Breathing (DB)** – Slow, deep "belly" breathing

- **Signal Breath (SB)** – One deep breath that serves as a signal to let go of tension

- **Quieting-Response (Q-R)** – An extremely short exercise designed to break up the stress response

III. Meditation

- **Focused Concentration (FC)** – Increasing awareness by attending to an object or image

- **Blank Slate (BS)** – Unstructured quieting

- **Structured Meditation (SM)** – A variety of images including religious and spiritual

IV. Cognitive Strategies

- **Thought Stopping (TS)** – Breaking up negative thoughts by saying STOP or visualizing an image that would do the same

- **Cognitive Restructuring (CR)** – Challenging and replacing unproductive thoughts with more productive ones

  ❶ Neutralizing "awfulizing" thoughts
  ❷ Eliminating "catastrophic" thoughts
  ❸ Avoiding "shoulds" and "oughts" as well as unfair comparisons and worry
  ❹ Eliminating unnecessary perfectionism and competitiveness
  ❺ Detecting negative self-talk
  ❻ Developing positive self-talk

- **Self-Health Programming (SHP)** – Affirmative suggestions

  ❶ Positive self-statements
  ❷ Repetition
  ❸ Gradual and positive results

V. Visualization and Imagery

- **Pleasant Scene (PS)** – Controlled daydream that transports you back to a place without worries. Intended as a brief strategy

- **Coping Imagery (CI)** – Imagine yourself handling a situation

- **Process Imagery (PI)** – Imagine your body's defenses and/or treatment fighting cancer (e.g. white blood cells as sharks)

- **Hypnosis and Self-Hypnosis**

VI. Behavioral and Environmental Strategies

- Cut down or eliminate stress-producing consumptive behaviors such as caffeine, sugar, and alcohol.

- Increase good nutrition

- Restful and Adequate Sleep

- Moderate Physical Exercise or Discharge

- Herbal and Naturopathic Remedies

- Aromatherapy

- Recreational Activities

- Humor and Fun

- Time Management

  ❶ Eliminate procrastination
  ❷ Analyze behavior before and after doctor and treatment appointments
  ❸ Structure time
  ❹ Add "padding" (extra time between tasks for delays and hassles)

- Build Communication Skills

  ❶ Listening – Active, involved
  ❷ Empathy – Putting your self in someone else's place
  ❸ Assertiveness

    a. Direct expression of feelings
    b. Self-enhancing rather than attacking
  ❹ Constructive Feedback – Effective criticism and compliments
    a. Be specific – Observational rather than judgmental
    b. Focus on behaviors not personality
    c. Tentative not dogmatic
    d. Avoid overloading at one time

- Emotional Discharge – Venting, Journaling, Therapy, Rituals

- Build a Support System

  1. Join a support group
  2. Improve existing relationships
  3. Reach out

- Seek Information

  1. Challenge assumptions
  2. Locate information resources and access them

- Improve Problem Solving Ability

  1. Identify and specify problem
  2. Identify contributing factors
  3. Brainstorm possible solutions
  4. Evaluate solutions for best positive outcome, short and long term consequences, and likelihood of follow through

- Change environmental stressors where possible

  1. Beautify surroundings

- Increase spiritual connectedness and sense of life's meaning

  1. Prayer
  2. Gratitude journal

GENERAL RELAXATION EXERCISE

(A Home Relaxation Exercise That You Can Tape)

Now, you are going to spend the next few minutes for the betterment of your health and well-being by relaxing and quieting your body—freeing it of all cares, worries, tension, anxiety and discomfort that you may be experiencing.

First, assume a comfortable position. It could be lying down or sitting up. Allow your posture to be open and relaxed. Gently close your eyes and keep your body still. Now, begin by taking a moderate breath into your lungs and as you exhale, just quietly think to yourself – *letting go*. As you do so, feel all the tensions, pressures, and a strain beginning to flow out of the body as the breath is exhaled from the lungs. Okay, now take a second moderate breath into your lungs and as you exhale again think quietly to yourself—*letting go*. Feel your body becoming more comfortable and quiet. Now, a third moderate breath and this time as you exhale think to yourself – *I am relaxed*.

Now, allow your breathing to slow down. Breathe deeply, but comfortably. Take in the breath gently through your nostrils and let your diaphragm rise instead of your chest. Let your breathing become smooth, with no pause between exhaling and inhaling. Notice that with each breath you take, you become more and more relaxed and as you exhale, you breathe out all frustration, tension, and discomfort. Focus your attention on the space in the middle of your forehead. Continue breathing slowly and smoothly. Just allow yourself to develop a comfortable and relaxing rhythm. If your mind wanders, gently bring it back to the space in the center of your forehead. Now just continue breathing in this fashion for the next several minutes...(3-minute pause).

Now allow your breathing to become automatic and concentrate your attention on the following phrases for quieting the body and the mind. Silently, repeat each phrase and then feel each area of your body relax as you just let go. Concentrate on the sensations you are feeling, as you become more and more relaxed.

> I am beginning to feel quite relaxed.
> I am beginning to feel quite relaxed.
> My feet feel warm, heavy, and comfortable.
> My ankles feel warm and quiet.
> My calves feel warm, relaxed, and comfortable.
> My knees feel quiet.
> My thighs feel warm, heavy, and relaxed.
> My hips, groin, and buttocks feel warm, relaxed, and quiet.
> The whole lower portion of my body feels warm, heavy, and relaxed.
> My stomach feels quiet, peaceful, and at ease.
> My chest feels light and it is easy to breathe.
> I can feel all the muscles in my back becoming loose and limp.
> My shoulders feel warm, quiet, and relaxed.

My arms feel warm, heavy, and comfortable.
My hands feel warm, heavy, and quiet.
My neck feels comfortable and relaxed.
My jaws, eyes, forehead, and scalp feel smooth and still.
I can feel all the muscles of my face becoming loose and limp.
Now, I will take a moment to relax any part of my body that still feels tense.
Now, that my body is relaxed, I will quiet my mind.
I withdraw myself from my surroundings and I become serene and still.
Deep within myself, I can see myself as relaxed and peaceful.
My mind is quiet.

Now, I will spend a few minutes experiencing and enjoying this state of total mind-body relaxation. (3-minute pause)

And now allow yourself a few moments in your mind's eye to develop an image of your self as you want to be – as the way you can be. See yourself becoming calmer and more relaxed. See yourself doing the activities you want to do, full of energy and the enjoyment of life. Appreciating your body and your self. (1-1/2 minute pause).

Now, if you wish to return to a state of full waking alertness – take two moderately deep breaths. Gradually open your eyes and gently stretch if you want. Feel refreshed and renewed.

Don't forget to congratulate yourself for having taken time to participate in your health and well-being.

BIOFEEDBACK

What is biofeedback? The word consists of two parts: *bio,* which refers to any biological process occurring, and *feedback,* which means returning information to a person about some aspect of behavior. The bathroom mirror is an example of feedback. When we look in the mirror to comb our hair, we use that visual information to style our hair the way we want. The bathroom scale is another example. We use the information from the scale to guide us to change our eating behavior.

Biofeedback extends this process under the skin to biological processes. Using the biofeedback equipment, we receive immediate information in some understandable form so we can develop mental strategies to exert greater control over physical processes. We are not usually conscious of these processes. For example: it is hard to tell whether our pulse is 80 or 90 beats a minute unless we stop and measure it.

Biofeedback uses extremely sensitive electronic instruments to measure a wide variety of physical processes like brain wave rhythms, muscle activity (tension), skin response, skin temperature, and blood pressure. There are other modalities of biofeedback, but those are the most commonly used. If we can measure a process, then we can feed it back to the person. These hard-to-sense internal biological activities are changed into a more easily sensed form such as a tone, a column of lights, a digital readout, or numbers on a meter. The machines do nothing to the person using them. They are merely recording activity as it takes place, and returning the information in an easily understandable form.

Biofeedback can help people learn to relax or regain control of certain physical functions that have been lost like leg paralysis after a stroke. Biofeedback is neither necessary nor helpful for some people, but for others who are very out-of-touch with their bodies, it may be the additional information that they need.

An example might be helpful at this point. Mrs. Jones gets very tense before chemotherapy treatments. She has not had a catheter, so she needs to have injections done for her chemotherapy. The nurse noticed that she tightens up her arm before her injections, so that the veins become constricted and it is harder to treat her. Dr. Edwards begins to do muscle (EMG, electromyography) biofeedback with her so she can learn what it feels like when she is beginning to tense her muscles. Dr. Edwards places small electrodes (sensors that measure electrical activity) on the skin surface of her forearm. Her feedback consists of a tone that speeds up and is a higher pitch when she tightens up (has more electrical activity) and the tone becomes lower in pitch and slows down (less electrical activity) when she relaxes. By becoming aware of changes that occur in her muscle tension, Mrs. Jones can learn what relaxation feels like without the equipment and call it forth when she goes to the clinic. This could make her treatments go more easily for both Mrs. Jones and the staff.

Muscle tension measured by EMG (electromyography) is probably the most common form of biofeedback. It is used with muscle tension headaches, anxiety and other tension disorders, pain problems, and intense fear problems (like needle phobias). Some health problems require increasing muscle tension, such as people who have been paralyzed by a stroke.

Skin temperature biofeedback is another common modality. A person can learn to increase or decrease body temperature in a certain area. A temperature increase means that more blood is flowing into that area and blood vessels are more dilated (open). This, too, could be useful when someone becomes tense before treatments. I have seen patients become so tense, they have actually pushed the needle back out of a vein. When people become stressed, the muscles tighten around and in the blood vessel walls and constrict blood flow to an area. This is often why people get cold hands and/or feet when stressed. Skin temperature biofeedback is useful in the treatment of migraine headache, Raynaud's Disease, a problem with circulation especially to the hands and feet, and other conditions with decreased circulation, like diabetes. Thomas Burish and his associates (see earlier citation) have used both EMG and Skin Temperature

Biofeedback with cancer patients. They have been helpful when used together with guided imagery or progressive muscle relaxation.

Skin response is another common biofeedback mode. When a person becomes emotionally aroused, there is greater sweat gland activity. This can be measured in a number of ways. The most useful measure is **Skin Conductance Response** (SCR), which detects how much of an electrical current is more easily carried along the skin. This has been associated with the lie detector test because it quickly changes when people are aroused. It is used to treat people who worry excessively or who stay chronically tense.

A fourth major type of biofeedback is work with brain wave rhythms. It is now commonly called **neurofeedback**. Sensors are placed on the scalp surface and the electrical activity from that brain area can be translated into a tone that increases or decreases with the amount of the activity being measured. Other types of displays can be done like a digital readout or a graph or visual design that can be increased or decreased with the brain wave activity being measured. Again, using the feedback a person can learn to selectively increase or decrease a brain wave rhythm.

There are several types of brain wave rhythms that are trained. The *alpha* wave is probably the most famous for inducing relaxation. *Theta* is used to help increase deeper meditative states and imagery. *Beta* waves are increased when there is an attention problem and *delta* when sleep is a difficult.

These types of biofeedback can be used with problems like insomnia, obsessive thinking, depression, and pain reduction. It has been successful with treating seizure (sensory motor response, SMR and ADD/ADHD disorders as well as substance abuse. As you can imagine, learning to control these brain rhythms takes a great deal of practice and may require forty to sixty or more one-hour sessions. For many patients, the results may well be worth the effort. Researchers and clinicians are again using this feedback with cancer patients. Gary J. Schummer, Ph.D. presented work at the 1995 Conference of the Society for Neuronal Regulation on the effects of increasing *alpha / theta* neurofeedback and *alpha* brain wave stimulation on HIV+ subjects. Their findings showed 31 percent and 34 percent increases respectively in the number of T-4 helper cells, which are an important component of our defense system. Conceivably, this type of biofeedback may be helpful with cancer patients trying to improve their bodies' defenses.

Biofeedback is an exciting area that shows mind-body interactions and how we can exert a more powerful influence over our body than previously believed. The process looks like the following diagram.

```
┌─────────────────┐      ┌─────────────┐      ☺
│ Stimuli Processed│──────│  Physical   │─────→
│ in Brain and     │      │ Regulation  │
│ Successful       │      │ Achieved    │
│ Strategy         │      └─────────────┘
│ Developed        │                          │
└─────────────────┘                           ↓
        │                          ┌──────────────┐
        ↓                          │  Electronic  │
┌─────────────┐                    │  Monitoring  │
│  External   │←───────────────────│    and       │
│  Display    │                    │  Processing  │
│  of         │                    └──────────────┘
│ Information │
└─────────────┘
```

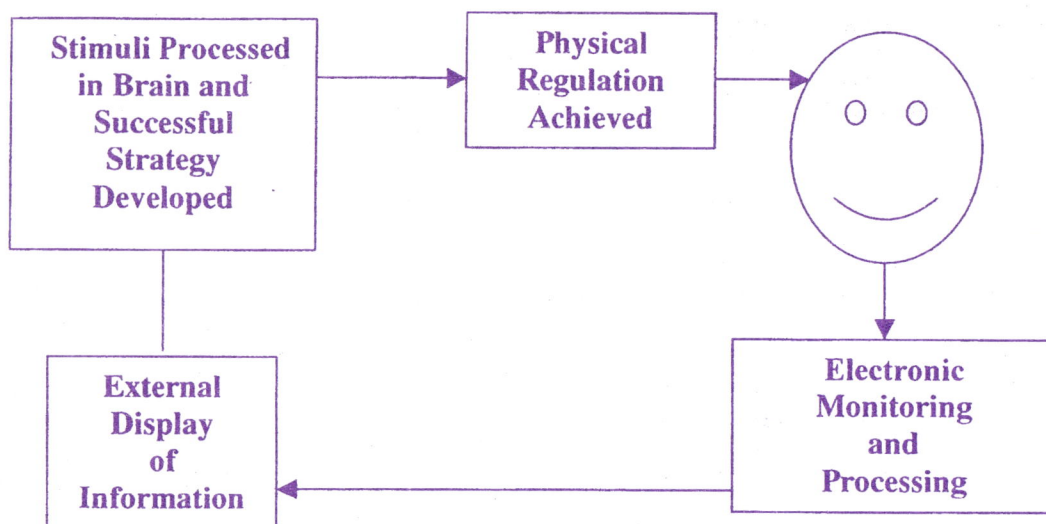

Some Important Points about Biofeedback:

❶ Different bodily processes can be monitored by using certain types of sensors.

❷ The person can be made aware of these internal processes by use of external signals such as a tone or lights.

❸ The machines do not do anything to the person. The person controls the internal processes and external signals, does the learning and acquires the skills.

❹ The machines merely speed up the learning process by making it easier for the person to feel what is going on inside his/her body.

❺ The eventual goal is for the person to learn what different states feel like and gain control of those states so that he/she can get off the machines and get into a particular state when necessary.

❻ Biofeedback can help treat many types of health problems. Biofeedback does not mean treatment done to you – it means active participation to learn skills for a lifetime.

❼ Biofeedback is not for everyone. Many of the same benefits can be derived from many of the other techniques discussed in this manual such as relaxation training and guided imagery.

FEELING PAUSE

The **Feeling Pause** is intended to help you become more familiar with how you are internally responding to the world around you. It is very easy for us to lose touch with how our bodies feel, as we stay focused on the external world. We can become used to being tense because that may be how we normally feel and we stop paying attention. Also, because stress often gradually increases throughout our daily living, it is difficult to detect. Review the earlier section on "**The Additive Nature of Stress.**"

By taking a few moments to monitor your feelings at several strategic points during the day, you will be better able to keep tabs on your stress level. Should your stress level be low, continue with present activities. Should it be moderate or high, it is then important to immediately engage in some activity no matter how brief, to reduce your tension level. The best way to handle stress is to continually keep it manageable by listening to your body, as you do with the **Feeling Pause**. You can almost always stay on top of your stress. Just remember these primary points.

❶ From time to time throughout your waking hours (every 15 – 20 minutes if possible, but no more than 30 minutes), take a few moments to get in touch with your tension level.

❷ Just pause for a moment and close your eyes if you are able.

❸ Focus on the area or areas of your body that appear most sensitive to stress (i.e., neck muscles, stomach, cold hands, etc.). This is your **stress barometer**. Use the **Stress Symptoms Checklist** that follows to help you identify what part(s) of the body serve as your stress barometer.

❹ Mentally record your stress level and write it down if possible. If it is a zero (for no tension), or a 1 (for mild tension), continue what you are doing. If it is a 2 (for moderate tension), or a 3 (for severe tension), let that be a signal that you are pressured or hassled and need to avail your self of some relaxation strategy.

STRESS SYMPTOMS CHECKLIST

Morry Edwards, Ph.D.

Even though there is a general response to stress, each person responds somewhat differently. In order to learn to control stress, the first step is to become aware of what physically happens to you when stressed. Please respond to each stress symptom by marking the number in the appropriate box and adding up the numbers for each subtotal. Note that this is an awareness tool and not intended for diagnosis.

	Never 0	Rarely 1	Sometimes 2	Often 3
MUSCULAR Tension Headaches				
Teeth Clinching				
Muscle Spasms Location:_____				
Muscle Tightness Location:_____				
Muscle Shaking or Tremor Location:_____				
SUBTOTAL				
RESPIRATORY Inability to Breathe				
Hyperventilation				
Chest Tightness				
Holding Breath				
SUBTOTAL				

	Never 0	Rarely 1	Sometimes 2	Often 3
CARDIO-VASCULAR Irregular Heartbeat				
Racing Heart				
Migraine Headaches				
Cold Hands or Feet				
Blushing				
Dizziness				
Faintness or Fainting				
SUBTOTAL				
GASTRO-INTESTINAL Diarrhea				
Constipation				
Nausea				
Vomiting				
Gas				
Abdominal Pain				
Stomach Butterflies				
Ulcers				
Indigestion				
SUBTOTAL				

	Never 0	Rarely 1	Sometimes 2	Often 3
EMOTIONAL Obsessions, Unwanted Thoughts				
Low Self-Confidence				
Feeling Fearful				
Depression				
Generalized Anxiety				
Irritability				
SUBTOTAL				
BEHAVIORAL Fatigue				
Sleep Difficulties				
Eating Too Much or Too Little (Circle Which)				
Accidents				
Smoking				
Sexual Problems				
Nail-Biting				
SUBTOTAL				

	Never 0	Rarely 1	Sometimes 2	Often 3
OTHER (Specify) Rash				
SUBTOTAL				
GRAND TOTAL				

CHARTING FORM

Sometimes it is useful to keep track of feelings and symptoms. By keeping track of these, you can more easily notice changes, improvements, or patterns that may help resolve some problems. You can duplicate the form below as much as you need and use it on a daily, weekly, or monthly basis. Use it like a graph and color in the boxes with different colors based on how little or much a percentage of the factor is experienced during that time period.

Date or Time Period: _____

	Health	Worry	Sadness	Hope	Anger
100%					
90%					
80%					
70%					
60%					
50%					
40%					
30%					
20%					
10%					

	Sleep	Interest	Pain	Energy	Joy
100%					
90%					
80%					
70%					
60%					
50%					
40%					
30%					
20%					
10%					

ACTIVITY FORM

Often the experience of cancer is tiring both emotionally and physically. The disease and its treatment can be quite stressful and fatiguing. This often causes frustration because people cannot do as much as they would like. Reconnecting with life affirming activities is vital! It remains important to be as active as you can without overdoing. Here is a form you can use to keep track of your activities during your waking hours.

Date:	Activity	Comments/ Energy Level
7:00 am		
7:30 am		
8:00 am		
8:30 am		
9:00 am		
9:30 am		
10:00 am		
10:30 am		
11:00 am		
11:30 am		
12:00 pm		
12:30 pm		
1:00 pm		
1:30 pm		
2:00 pm		
2:30 pm		
3:00 pm		
3:30 pm		
4:00 pm		
4:30 pm		
5:00 pm		
5:30 pm		
6:00 pm		
Evening:		

LAB RESULTS

Awaiting lab test results can be one of the most stressful experiences for a cancer patient. Some people feel more in control and less stressed by keeping track of these findings. You may notice patterns or changes that you can make. Here is a convenient form to use to keep a record. The clinic where you go may have a computer report in a form you can understand. They may be more than willing to make a copy, if you ask. If it is stressful for you to keep track, then it would be better not to chart results. You may copy this page as much as you like.

Lab Test	Date:	Date:	Date:	Date:	Date:

BREATHING

Aside from very obvious reasons, breathing is extremely important to our state of mind. Everyone I have ever worked with has been breathing and they breathe wherever they go. Breathing is an extremely portable relaxation strategy. People are often amazed at how calm they feel after just two minutes of focused, relaxed, diaphragmatic breathing.

Often when we are stressed, our breathing pattern becomes unconsciously labored in one of two patterns. We either breathe shallowly, quickly, and irregularly, so we come close to or actually do hyperventilate. This causes inefficient oxygen exchange. Or we hold our breath and tighten our upper body, especially to brace against the perceived threat. This too causes inefficient oxygen to be circulated and we again increase our anxiety.

Cancer can invade the lungs and make it very difficult to breathe. If that is a problem for you, other relaxation techniques that focus more on mental imagery may be helpful. Sometimes cancer is not the problem and people have a great deal of difficulty focusing on their breathing. They try too hard to control it or they do not like to focus on their bodies. Again, benefit may be obtained by using a different relaxation strategy, such as **The Pleasant Scene** (see the section on **Imagery**) or **The Quieting Response.** Remember: do not try to force these techniques. That will only succeed in causing you more frustration.

Deliberately slowing down our breathing can calm us right down. There are many variations of deep, relaxed breathing that can slow us down as well as energize us. There are two main exercises that will be discussed in this manual: **The Signal Breath** and **Diaphragmatic Breathing**.

The Signal Breath is a single breath that can begin to serve as a signal to let go of tension and calm down. This technique has three simple steps and is intended to be done anywhere, except while driving. Since it takes so little time, it can be done quite frequently, especially during waits for doctor appointments or tests.

First, you gently look up at the center of your forehead (right above your eyebrows). This eye movement should be a gentle motion and your eyes do not have to roll back in your head.

Second, while still looking up at the center of your forehead, begin breathing in, preferably through the nose. This inhalation should be smooth, quiet, and moderately deep. As you are breathing in, let your eyelids gently close.

The third step is when you reach the end of your inhalation, hold your breath for approximately two seconds and then using your lips as a valve, slowly and smoothly breathe all the way out until your gut feels empty. If you do this properly you will feel a pleasant tingly or floating sensation in your hands or feet or perhaps throughout your entire body. With practice this breath or even just the eye movement will begin to train you to let go. This technique is summarized on the following page.

A technique with a number of mental variations is **Diaphragmatic Breathing**. While the breathing part remains the same, what you do to keep your mind focused may vary. Some people just like to quietly sit and focus on their breath as it gently flows in and out. They may see the breath as one color (peaceful blue) going into the body and another color (tense red) flowing out. Some people count as they breathe in and repeat the count as they breathe out. Some suggest counting twice as long on the exhalation so the body will naturally tap into the relaxation response (the parasympathetic system). I typically have patients repeat a reinforcing phrase: on the inhalation "I AM" and on the exhalation "RELAXED." Another variation is to imagine peace coming into the body and worries flowing out. You can repeat "PEACE IN" as you inhale and "WORRIES OUT" as you exhale.

Whatever mental technique you use to keep your mind focused, your breathing should be slow, quiet, smooth and moderately deep into your diaphragm area. It is best to breathe in and out through the nose, allowing yourself to fall into a comfortable and easy rhythm. Breathe in fully without straining. Many people "over-breathe" and exert too much energy in their breathing. Do not force the breath. Instead let your breathing slow down gradually. When you reach the height of breathing in, slowly let the breath back out so it is like breathing in a circle. Start out for as long as you can and build up to at least five minutes. Take mini-relaxation breaks throughout the day as needed, when your body or mind let you know you're stressed (Remember **The Feeling Pause**). It won't change the world, but it will sure calm you down and enhance your ability to deal with things going on in the world. An example of the Diaphragmatic Breathing technique is provided in **A Relaxed Breathing Exercise**.

SIGNAL BREATH

Breathing is the most effective and most transportable relaxation technique available to us. We just need to become conscious of what we are doing because we often develop bad breathing habits and inefficient oxygen intake.

❶   Gaze upward so that your eyes are focused on the center of your forehead.

❷   Inhale slowly and evenly through your nose as you close your eyes. Hold your breath momentarily.

❸   Now exhale fully through your mouth. Empty the breath totally so that you feel it in your gut. Feel your self float or feel a warm tingly sensation in your hands and/or feet.

A RELAXED BREATHING EXERCISE

❶ Assume a comfortable position. It may be either sitting or reclining. Keep your posture open and relaxed.

❷ Gently close your eyes and focus your attention on the center of your forehead. Should your attention wander, gently bring it back to this space.

❸ Keep your lips closed and breathe through your nose. Let your breathing gradually slow down. Make sure that your inhalation is slow, smooth, and even. Allow your breathing to settle in a comfortable rhythm and breathe into your stomach area instead of your chest. Avoid breathing too slowly or too deeply. Let your breathing slow, without holding your breath. Picture it as a circle. When you come to the height of your inhalation, gently let the breath flow out evenly and slowly.

❹ As you breathe in, say to yourself, "I AM." As you breathe out say to yourself, "RELAXED." Keep repeating during the entire exercise.

❺ Repeat this exercise at convenient intervals at least four times for approximately five minutes each time. It can be done more frequently and for longer periods. The important aspect is to start listening to your body and start quieting down regularly throughout the day – **every day**.

QUIETING REFLEX (Q-R)
Developed by Charles and Elizabeth Stroebel

Quieting Reflex Training® is a program that involves a number of stress awareness and management techniques. The Stroebels developed this training program after noticing that many people who learned longer forms of relaxation did not use them more than three months even when they felt improvement. The Stroebels therefore, taught their patients both "long" twenty to thirty-minute relaxation exercises, and brief strategies like the **Q-R**. This exercise could be done throughout the day, when people would notice their tension increasing. It made the relaxation response more transportable and allowed people to "downshift" in "the heat of the moment," when they needed it most. This is a technique that can be used with **The Feeling Pause**. Once you identify sensations that indicate an increase in stress, follow the steps below. It may help you to routinely do the **Q-R** every few minutes. This practice will retrain you to substitute this relaxation response for your stress response.

❶ Smile – both inwardly and outwardly.

❷ Silently state a Quieting Phrase to yourself, such as, "I can handle this," or "Calm down."

❸ Take two comfortable, moderately deep breaths from the stomach area.

❹ Focus on your most tense or restless area(s) and let it loose.

❺ Repeat as often as possible – 50, 100, 200 times daily, especially when confronted with all those irritants of daily life. This may seem like a high number, but realize you are taking time to learn how to relax instead of being stressed.

❻ With adequate practice, the **Quieting-Response** will become automatic in three to six months.

HOW THINKING CAN WORSEN STRESS OR RELIEVE IT

There are two main ways in which we think. One way is verbally, in which we process our world by using **words**. The other process is non-verbal where we "see" **images** or use any of the other senses in our imagination to experience a situation. First, I will discuss thinking as a verbal process, which we will refer to as **self-talk**. Then I will discuss **imagery,** or as it is sometimes called, **visualization**.

SELF-TALK

While growing up, you probably heard the in-jest warning: "If you talk to yourself, you must be crazy!" Sometimes, it might have even been more specific as in: "If you answer yourself, then you are definitely crazy!" In actuality, we talk to ourselves a great deal of the time and we answer or argue with ourselves frequently as well. We call that process "thinking" and it is not an abnormal process, unless we are severely distorting what is really happening in a situation.

Self-statements or self-talk are called thoughts. These are also called **cognitions**, and play a major role in how we feel, behave, and physically respond. The main reason being that regardless of how accurate or distorted our self-talk is, we believe what we are saying to ourselves. This presents no problem, when we are accurately assessing a situation. However, when we are inaccurately "catastrophizing" or "awfulizing" a situation, it may cause us to become needlessly more anxious or upset, more inhibited, or to act impulsively. An example may be helpful at this point:

After being diagnosed with cancer, have you said something like this when you've felt a new ache or pain:

"I'm sure the cancer has returned or is spreading!"

There is no question that this kind of thought is very scary. People interpret a new ache or pain or even passing physical sensation as their cancer spreading or returning. While that may be a possibility, it is not a fact.

The first thing I would recommend is do not panic or get swept into fear. Analyze the symptom. If your neck hurts when you wake one morning, your immediate response may be to call the doctor. Instead, ask yourself, did I sleep in an awkward position last night? That might lead you to do some stretching or relaxation. The pain could then decrease without adding anxiety. If the pain persists, it may then be something to consult your doctor about.

Using the diagram on the following page, let's look at the process more deeply. In any situation we experience sensations. Sensations may come from the external world or from inside our bodies. They are then processed and become perceptions. Perceptions are different from sensations in that our brain acts as a filter based on past experiences. Our past experiences or learning history colors our way of looking at things. This leads to particular self-talk or thoughts whether we are conscious of what we are telling ourselves or not. Thoughts lead to feelings. Whenever we have thoughts about a situation, some of them are accurate and objective, while some of them may be inaccurate and distort things greatly. If our thinking is accurate, it often leads to appropriate feelings, which may not always feel good, but as it implies, are realistic given the situation. If we accurately perceive the situation and our feelings are appropriate, then eventually that leads to coping behavior and/or acceptance. Sometimes we have accurate thoughts about a situation, but we react disproportionately because we feel intensely about the situation. That can lead to denial or problems with behaviors, which are inappropriate. Lastly, we can have inaccurate thoughts about a situation that can cause us needless anxiety or depression, which can also lead to denial or maladaptive behavior.

Let's look at the example in the diagram:

"I'm going to die from this disease soon."

First, it may be important to check that assumption with your doctor to make sure it is an accurate assumption. Your doctor may tell you that it is an inaccurate thought because he/she has no physical evidence for that conclusion. Hearing that would provide reassurance, so needless stress could be avoided. But, what if it is an accurate thought? Then it would be appropriate to feel sad, scared, or angry. Working through that would be important so that you decide to live your life fully, possibly take trips or heal other areas of your life before dying. Sometimes, people do not get beyond feeling cheated or resentful and may do something inappropriate like establish a huge credit card debt to leave their family.

Real Life Situation

Sensations

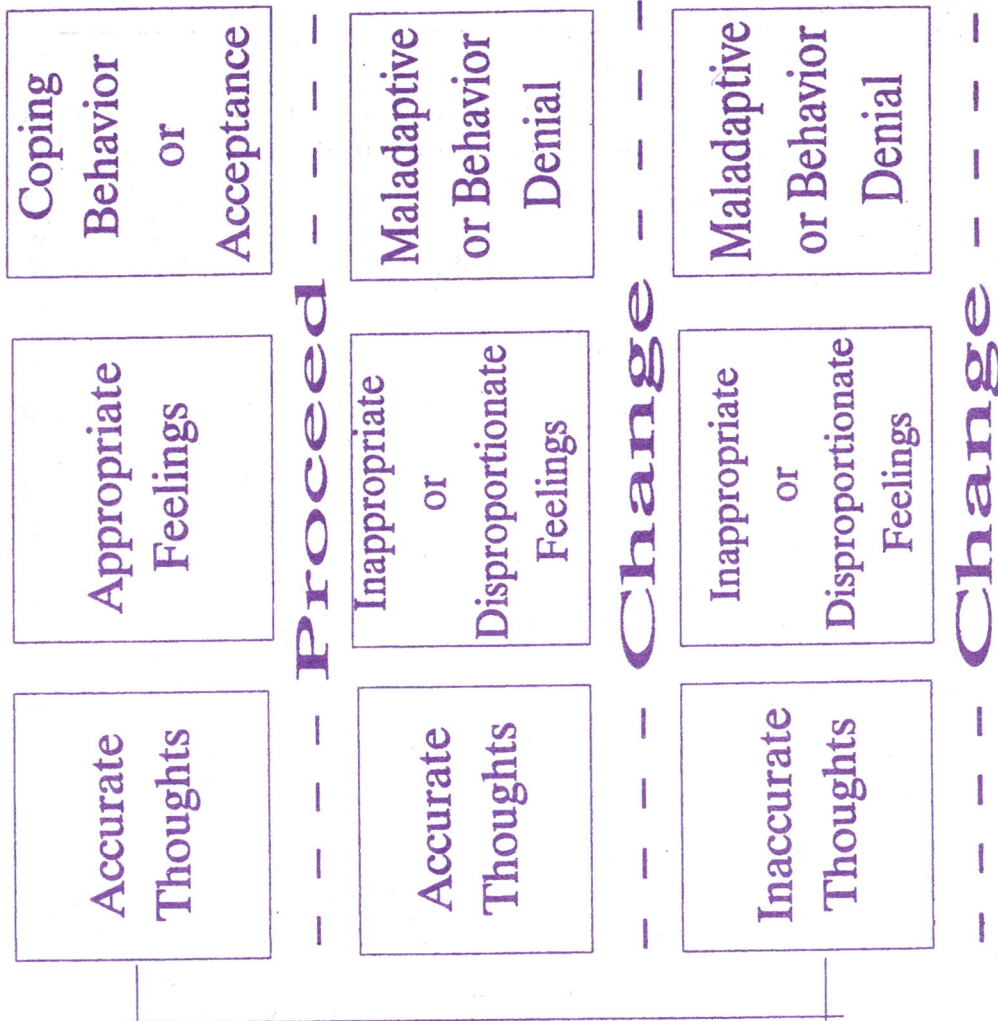

Perception + Processing

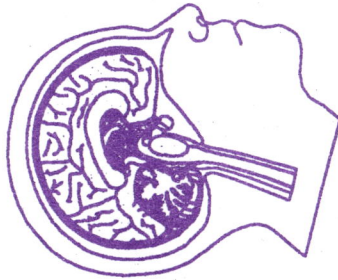

Accurate Thoughts	Appropriate Feelings	Coping Behavior or Acceptance

Proceed

Accurate Thoughts	Inappropriate or Disproportionate Feelings	Maladaptive or Behavior Denial

Change

Inaccurate Thoughts	Inappropriate or Disproportionate Feelings	Maladaptive or Behavior Denial

Change

Thought

I'm going to die from this disease, soon.

Check Assumptions with Doctor

I may die from this, but I can still live fully until that happens.

Because so much of our thinking becomes a habit, it becomes automatic or subconscious. That is, we don't realize what we are telling ourselves. We may need to exert effort to become conscious of our thinking, so that we can challenge our inaccurate thoughts and change them so it does not lead to unproductive and needless emotional turmoil. The worksheets on **Steps to Thought Restructuring** and the **Self-Statement Inventory for Cancer Patients** can help you with that process.

Two other important techniques that involve self-talk are **Autogenic Training (AT)** and **Affirmations**. Autogenic means "self-generated" and this refers to self-generated phrases that we repeat silently or out loud. Dr. J. H. Schultz and Dr. W. Luthe found that the repetition of certain phrases could produce physical changes. For example: The use of the phrase, "My hands are warm and heavy" increased blood flow to the hands and therefore raised their temperature. This passive relaxation technique could be used to decrease blood pressure or help prevent migraine headaches.

Affirmations are positive self-statements. People can repeat certain phrases to help change beliefs that they have, especially negative or self-defeating ones. Changing attitudes or deep-seated beliefs takes time, so it requires some patience. Facing a disease like cancer can lead to important revelations. We've often trained ourselves to be patient with our children, our aging parents and other co-workers or friends. If you have not mastered it before, now is the time to learn to practice patience yourself! It also may seem artificial and unbelievable at first. Stick with it for at least a month. Surround yourself with the phrase. Put it on your mirror in the bathroom, on your computer screen saver or anywhere else where you are likely to see it and repeat it. Some examples you may want to use are:

I will cultivate peace and not worry about things I cannot change.

I will be totally involved in each moment that I live.

My immune system is strong and effective.

My body is continuously healing itself.

I know that my innermost being is magnificent, wise, and loving.

I am responsible for my thoughts and actions.

I release all anger, fear, and depression.

I forgive myself for all past transgressions.

I forgive others for their transgressions.

List some of your own affirmations:

_____

_____

_____

STEPS TO THOUGHT RESTRUCTURING

❶   Pick a situation that causes you discomfort or stress. Close your eyes and imagine yourself in that situation. Feel what happens to your body and listen to what kinds of thoughts come into your head.

❷   Identify and isolate individual thoughts in that situation and write them down so you can see exactly what you are saying to yourself.

❸   Analyze each thought that you write down as accurate (if it is objectively verified by the situation) or as inaccurate (if it is a subjective distortion such as an overgeneralization of that situation).

> **Example**:   *"They had some problems treating me this time."* **(Accurate Thought)**

> **Example**:   *"Nothing goes right for me when I come to the clinic."* **(Inaccurate Thought)**

❹   If thoughts are accurate, make sure the feelings and behaviors are appropriate, proportional to the event, and cope in a constructive way.

If thoughts are inaccurate, then develop a thought that is more accurate, reassuring, or calming and substitute that thought each time you hear yourself say that inaccurate thought. At the very least, stop repeating the negative or anxiety-producing thought.

> **Inaccurate:**   *"This chemotherapy just makes me sick."*
> **Accurate:**   *"Even though I get sick, the tests show it is reducing the tumor."*

Try using the **Thought Restructuring Worksheet** on the next page.
You may copy that form as much as you wish.

THOUGHT RESTRUCTURING WORKSHEET

Situation:

_____

_____

_____

_____

Current Thought	Accurate (Yes / No)	Coping Thought

SELF-STATEMENT INVENTORY FOR CANCER PATIENTS
Morry Edwards, Ph.D.

Below are listed a number of statements that people make to themselves while undergoing medical procedures such as chemotherapy. Please read each self-statement and indicate how often you might think it during your typical chemotherapy treatment. After you have carefully read each item, please circle the appropriate number that relates to your thoughts.

Name_____ Sex_____ Date_____

Diagnosis_____ Age_____

Time since diagnosis (approximately)_____Years _____Months

Time since starting chemotherapy _____ Years _____Months

Frequency of treatment _____

Chemotherapy drugs_____

		Hardly Ever		Sometimes		Very Often
1.	I was thinking the procedure could save my life.	1	2	3	4	5
2.	I was thinking about the wonders of medical science and how lucky I was that they could do this for me.	1	2	3	4	5
3.	I was thinking how helpful my treatment is.	1	2	3	4	5
4.	I was concerned that the nurses and doctors were negative and incompetent.	1	2	3	4	5
5.	I kept thinking the treatment was painful and hard to get through.	1	2	3	4	5
6.	I was worried about the bad things the staff said might happen.	1	2	3	4	5
7.	I was feeling confident in the skills of the doctors and nurses.	1	2	3	4	5
8.	I was listening and expected them to say something bad about my health.	1	2	3	4	5

Self-Statement Inventory for Cancer Patients, *Morry Edwards, Ph.D.* (continued)

		Hardly Ever	Sometimes			Very Often
9.	I kept thinking that the treatment would cause complications that would never go away.	1	2	3	4	5
10.	I kept reminding myself to just think about pleasant things and take my mind off the procedures.	1	2	3	4	5
11.	I kept reminding myself about all the times in the past when I had been successful in coping with stress and pain and this was not any worse than those situations.	1	2	3	4	5
12.	Since the treatment was not discomforting, I was thinking about other things.	1	2	3	4	5
13.	I kept worrying the treatment would kill me.	1	2	3	4	5
14.	I kept expecting the treatment to damage my body.	1	2	3	4	5
15.	I kept thinking how I did not really want to have chemotherapy.	1	2	3	4	5
16.	I was thinking about the things that I need to do to be a good patient.	1	2	3	4	5
17.	I kept thinking how much I disliked the smells of the clinic.	1	2	3	4	5
18.	I kept thinking how sick I would get after the treatment.	1	2	3	4	5
19.	The staff and the facilities made me feel confident about what was going to happen to me.	1	2	3	4	5
20.	I was thinking how much easier my treatment would go, if I kept myself relaxed or focused on pleasant thoughts.	1	2	3	4	5

Adapted from Kendall, Williams, Pechacek, Graham, Shisslah and Herzoff's Self-Statement Inventory for Heart Catheterization.

IMAGERY

The second way that we process and remember is non-verbally with images or pictures. When you close your eyes and concentrate for a moment, you may be able to "see" past events, daydream about fantasies, picture people's faces or even anticipate future events. This "seeing with the mind's eye" is called visualization and the process is called imagery. Visualization and imagery are terms that are used interchangeably. Visualization is more limited to our sense of sight, while imagery includes all of our senses – sight, hearing (auditory), taste, smell (olfactory), touch, as well as movement (kinesthetia) and inner sensation (proprioception). Certain feelings can be recaptured through imagery as we remember the secure feeling of smelling freshly baked apple pie or the pleasure of tasting our favorite food. We'll use the term imagery to mean experiencing or sensing an event with the mind at one's own direction.

Your imagery is important because it puts you in touch with how you feel about objects, people, process, events, and goals. When you create a concrete image, you can better detect and examine your deeper feelings. For example, if you were to view your cancer as a big, furry beast, it may show that you really feel that it is powerful and hard to fight. If your cancer seems like a pool of stagnant water, you may feel better able to drain that pool dry.

Imagery is important because awareness of your true feelings will give you the basis from which to change or reinforce your attitude. You may not be able to change your diagnosis, but you can change your attitudes and beliefs about that diagnosis. Changing the way you view your disease may make you feel more positive about healing from it. Working with imagery can change your outlook and influence the way your body fights disease and responds to treatment. Our beliefs and attitudes can affect the way we behave. If people feel hopeless, they tend to give up and stop trying to get better. If they feel hopeful, no matter what form that hope takes, they tend to keep trying. Holding on to some form of hope helped people survive concentration camps. It can help you better handle your cancer treatments and get well.

Different people prefer and process information better in one sense over another. For instance, some people may be able to close their eyes and see very vivid and detailed "pictures" of a person's face or a pleasant place they visited quite some time ago. Some people close their eyes and can't see a thing! They may, on the other hand, be able to hear a fantastically intricate melody in their mind just fine. Researchers have determined that the same nerve pathways are activated when we image the sense as when we are actually using it. More than fifty years ago physiologist, Edmund Jacobson, showed that when a person imagines running, small but measurable changes occur in the muscles that would be involved. Many athletes use this method to improve their performances. To assess which sense you process the best, try this simple exercise below.

IMAGERY EXERCISE

Try to imagine five common things in each of your five primary senses and rate how vivid each image is.

Use a 0-3 scale to rate each item:

0 = nothing at all
1 = slightly clear and detailed
2 = moderately clear and detailed
3 = extremely clear and detailed.

Examples may be: the smell of apple pie, your favorite song,
a family member's face, the taste of mint, or the warmth of the sun
on your skin.

Based on your experience with the above exercise, you may want to first practice imagery with your best sense. This would reduce frustration and enable you to better establish a routine. Using as many senses as possible is the best way to actually experience what you are imaging. To enhance your imagery even further, Jeanne Achterberg, Ph.D., Barbara Dossey, R.N., M.S., FAAN., and Leslie Kolkmeier, R.N., Med. in their book, *Rituals of Healing*, suggest **cross-sensing** or **synesthesia.** Here you imagine using different sensory modalities like seeing the sounds of your muscles loosening or hearing the color of your relaxation.

If you have a great deal of difficulty imaging with your mind's eye, then begin with a more concrete practice. Try concentrating on a specific object with your eyes open. You should start with something simple like a candle, a piece of fruit, a flower, or a basic colorful geometric shape. Gently stare at the object and take it in fully. Just try and observe as much detail as you can. After a few minutes of this, close your eyelids and you may see an "afterimage." After practicing this activity routinely, you should be able to "see" more and more clear "afterimages." Focusing on an image like this helps quiet the mind. It may particularly help you if your mind is racing with anxious thoughts. Instead of trying to shut your thoughts out, it is a better strategy to fill your mind with an image so the racing thoughts are pushed out without effort.

Another concrete way to assist your imagination is with the use of posters or art. A pleasing picture of a relaxing scene or a work of art that you enjoy can serve as a way to focus your attention and slow your mind and body down. You could also get some pictures of your immune system getting rid of cancer cells and use that with the more sophisticated imagery discussed later in this section.

There are different types of imagery, which may be put to specific uses. We can use imagery to relax such as doing the **Pleasant Scene** where we take ourselves back to a special place, which was very calming. This could be lying by the ocean (with appropriate sunscreen of course!), sitting in a meadow of flowers, or on a mountaintop, or taking a slow walk in the woods. If you were relaxed when you were in that situation, then reliving it should trigger the same relaxing sensations.

We can use **Coping Imagery** to practice learning constructive ways of reacting to stressful situations. Athletes use this type of imagery to improve their performance by practicing and mentally correcting what they have been doing wrong. Research has shown that when we imagine ourselves in certain situations, our muscles slightly react to what we are thinking. So mentally practicing your golf swing correctly actually can help you play better.

Remember if you are "seeing" negative pictures in your mind, then you are likely to become more anxious. If you are seeing positive pictures, then you are more likely to feel calm (or energized depending on the situation) and confident in handling them in real life. For example: If you see the nursing staff having difficulty starting your chemotherapy, you could become anxious and tense up for your treatments which might make your nurse have more difficulty, thereby causing you to tense up even more and so on. By doing the imagery and instead "seeing" your treatment as going very smoothly and you being very calm in the situation, you would in turn be calmer in the situation and make it easier for you to be treated.

Having cancer makes us feel out of control and because our society has such a negative concept of cancer, we can feel helpless in overcoming it. When you use **Process Imagery**, you experience obtaining a goal one step at a time.

An example of this is the **Relaxation and Visualization Exercise to Enhance Immune Function**. In this practice, you imagine your natural killer cells increase and demolish cancer cells until your tumor is completely gone. You do not need to use violent imagery in this exercise. Many people prefer images of natural healing such as sunlight melting snowflakes.

A SUMMARY OF IMAGERY BENEFITS

> ➤ Imagery will decrease fear, tension, and pain.

> ➤ Imagery can improve your attitude by making it more positive and strengthening your "will to live."

> ➤ Imagery will help you affect bodily changes that can ease the treatment procedures and possibly make them more effective.

> ➤ Imagery may help alter feelings of helplessness and lack of control to confidence and ability to gain control events.

SOME HINTS FOR IMPROVING YOUR PROCESS IMAGERY

❶ Learn a basic relaxation technique first. This strategy may be a focused breathing exercise or a deep muscle relaxation or a pleasant scene. Let yourself slow down gradually. Repeat the phrase "letting go" as you focus on each muscle group or organ throughout your body.

❷ After allowing yourself to enter a peaceful state, begin to let your mind focus on a picture of your cancer. This image can be real or symbolic, but what is most important is that you picture the cancer as helpless and weak, vulnerable and easily defeated by your body's natural defenses and your medical treatment.

❸ Picture your body's defenses as strong, aggressive, numerous, and smart. They enjoy devouring cancer cells, which are no match for them. They easily destroy the cancerous cells and then help flush them from the body.

❹ See your treatment as a strong and effective ally against weak cancer cells. It is important to see your treatment in a positive way rather than something you have to endure.

❺ Now begin to imagine your cancer shrinking due to the combined efforts of your own natural defenses and the treatment you are receiving. Just imagine the white blood cells and chemotherapy and/or radiation just tearing into cancer cells and wiping them out. Continue until your cancer has totally disappeared. As it shrinks you get stronger and healthier. Feel your body become more and more powerful as it becomes well again. Imagine healing and harmony as the end result.

❻ See yourself as well and reaching your goals in life. The more meaningful your life is and the stronger your reasons for getting well, the better you may do.

❼ Praise yourself for participating in your health and well-being. Try to practice as regularly as you can and approach your practice as something positive. Also use imagery that is emotionally powerful and believable. Use images that are compatible with your view of life and that come from deep experiences you have observed. This should not be just an intellectual exercise.

RELAXATION AND IMAGERY HOME JOURNAL

You can use this form to record your practice and progress of relaxation and imagery. I encourage you to modify this format if you desire. You may make additional copies of this form, if you like it. Simply record your tension level before and after your relaxation period (**0 = no tension or stress, 1 = mild, 2 = moderate, and 3 = severe,** the same as when you are doing the Feeling Pause). The conviction level relates to how vivid, real, or believable your imagery was (**0 = not at all, 1 = slightly, 2 = moderately, and 3 = extremely**). A space is provided to list any particular images or thoughts, including affirmations you had and how they felt, as well as a space for general comments.

Name _____ Date_____

Images and Thoughts:

_____

_____

_____

_____

_____

_____

	Not at All	Mild/Slight	Moderate	Severe/Extreme
Tension Level Before	0	1	2	3
Tension Level After	0	1	2	3
Conviction Level	0	1	2	3

Comments _____

_____

_____

PLEASANT SCENE

❶ Take a few minutes several times daily for this controlled daydream.

❷ Assume a comfortable position and gently close your eyes.

❸ Let your breathing slow down and your body "go loose."

❹ Allow your mind to quiet down and then drift back to a place or event in the past where you felt perfectly relaxed, calm, and at peace.

❺ Let this image come as vividly in your mind as possible. Involve as many of your senses as you can. Example: See the blue of the ocean. Feel the warmth of the sun on your body. Hear the rhythmic rolling of the waves and the sea gulls. Smell the salt in the air.

❻ To return to a state of full waking alertness, take two deep breaths and gently open your eyes after a few minutes.

COPING IMAGERY

Often, when a person is anticipating a specific stressful situation, he/she develops pictures as well as thoughts about the stressful situation. Usually, these add to the anxiety because they are negative, or the person sees him/herself not coping well with what is happening. An example is going for chemotherapy and imagining the staff having problems with you.

A way to offset this aspect of stress is to see yourself handling the situation in a calm and confident manner. Follow these simple steps repeatedly until you feel relaxed thinking about the stressful situation. Then you can gradually expose yourself to the real stressful situation.

1.   Get comfortable and relaxed.

2.   Think about a stressful scene.

3.   Begin to get worried.

4.   Switch from thinking about the stressful scene to relaxation.

5.   Allow yourself to relax again.

6.   Think about the scene with you coping effectively this time.

7.   Feel totally relaxed and calm with the stressful scene.

8.   Repeat until Step 7 is achieved.

SYSTEMATIC DESENSITIZATION (SD)

Systematic Desensitization is similar to **Coping Imagery** except that it gradually leads you from the least anxiety-provoking aspect of the situation up to the most anxiety-provoking part. Systematic Desensitization is a mental process and can be done in coordination with a relaxation exercise. *In vivo* **Desensitization** is where you gradually expose yourself live, while you are reconditioning yourself with relaxation. This is a technique that has been very successful with many types of phobias where a person has had an irrational fear of a specific situation or object (e.g. doctors' offices, a stressful medical test, needles, etc.).

The first thing to do is identify a situation that gives you an undue amount of stress. Break it into a hierarchy like the one shown on the next page. Rate each item of the hierarchy as to the amount of **Subjective Units of Distress (SUDS)** from 0 to 100 with 100 being the most stressful. Start out with the first item by using all your senses to imagine yourself in the situation. If you get anxious, practice a relaxation technique like slow breathing or imagining a pleasant scene until your SUDS rating approaches zero. Then go back to that scene. Keep alternating until you are able to imagine that stressful scene twice for at least 30 seconds (your estimate) without getting anxious. Then go on to the next item of your hierarchy until you can master the situation. The next step may be to go through things live. The more exposure to the situation you have, the more likely you will become desensitized to it.

SAMPLE HIERARCHY FOR OUT-PATIENT TREATMENT

SUDS: Subjective Units of Distress

- (20) Getting ready for clinic appointment
- (40) Driving or riding to clinic appointment
- (60) Arriving at clinic parking lot
- (75) Walking into clinic
- (80) Signing in at reception desk
- (90) Sitting in clinic waiting room
- (100) Sitting in chair beginning to receive chemotherapy

Now, if there is a particular situation that is giving you significant stress, make your own hierarchy with SUDS below.

Situation: _____

SUDS

- ( )_____

- ( )_____

- ( )_____

- ( )_____

- ( )_____

- ( )_____

- ( )_____

- ( )_____

- ( )_____

RELAXATION AND VISUALIZATION EXERCISE TO ENHANCE IMMUNE FUNCTION

This is an exercise that will help you relax and improve the function of your immune system. Take a moment and assume a comfortable position. You can either lie down or sit in a comfortable chair. Let your posture be relaxed and open. If you are seated, allow your arms to rest in your lap or on the arms of the chair. Keep your feet on the floor and avoid crossing your legs. If you are lying down, let your arms rest by your side and keep your legs out straight.

Focus your attention on the center of your forehead and allow your eyes to gently tilt up to that area. Now close your eyelids and let your mind begin to go blank. Begin slowing down your breathing by taking air in your nostrils and into the diaphragm area instead of your chest. Let your breathing become slow, smooth, and moderately deep so that you develop a comfortable rhythm. Do not hold your breath, but gently exhale when you come to the height of your inhalation. Imagine your breathing as a circle. When you breathe in, say to yourself, "I AM," and when you breathe out, say to yourself "RELAXED." Now just allow your breathing to fall into a slow, comfortable pattern and begin slowing down your body ... ...

Now that you have begun to slow down, imagine a wave of warm, pleasant relaxation begin to flow through your body. Begin at your forehead. Feel this wave of warm, pleasant relaxation flow across your forehead and around your eyes, around your eyes and down through your cheeks and jaws allowing all the muscles of your face to become loose and limp, relaxed and quiet. And feel this wave of warm, pleasant relaxation begin to flow into your neck, down your neck and across your shoulders. And feel this wave of warm, pleasant relaxation flow across your shoulders and into your arms, down your arms and into your hands making them feel warm and heavy.

Warm and heavy.
Warm and heavy.
Warmer and heavier.
Warmer and heavier.
Heavier and heavier.
Warmer and warmer.

And now feel this wave of warm, pleasant relaxation flow across your chest making it light and easy to breathe. And over your stomach making it quiet and peaceful. And now feel this wave of warm, pleasant relaxation begin to flow down your back so that you feel every muscle of your back become loose and limp, smooth and relaxed. And notice that the whole upper portion of your body feels warm and pleasant, relaxed and calm, quiet and peaceful. More and more relaxed.

99

More and more at peace. More and more quiet. More and more calm. And just allow this wave of warm, pleasant relaxation to begin flowing into the lower portion of your body. Around your hips, groin, and buttocks. And now feel this wave of warm, pleasant relaxation flow into your legs. First, your thighs. Then feel this wave of warm, pleasant relaxation flow into your knees and your calves and finally feel this warm, pleasant wave of relaxation flow into your feet making the whole lower portion of your body feel warm and heavy. Warm and heavy. Warmer and heavier. Heavier and warmer. Heavier and heavier. Warmer and warmer. Just allow all remaining tension to flow out through the feet.

Now, take a few moments to experience and enjoy this state of warm, pleasant relaxation. While you are allowing your mind and body to become calm, quiet and peaceful, repeat to yourself the RELAXATION WORD OF THE DAY. Today the RELAXATION WORD OF THE DAY is _____. (Calm, Quiet, Peace, Love, etc.)

Now that you are feeling relaxed and peaceful, remind yourself how wonderful your body truly is. Remind yourself that your body has countless cells that help you fight intruders. Your body's defense system continually patrols for any abnormal situation that requires attention. It may control an infection or spot an abnormal cell that needs to be destroyed and removed from the body. Let yourself imagine in any way that makes sense to you, these powerful helper cells that protect your body. It can be realistic or symbolic, but imagine this as vividly and powerfully and believably as you can. Imagine these cells as numerous, intelligent, and aggressive in doing their best to defend your body from harm and keep it balanced. Imagine these cells doing their job efficiently and effectively, keeping harmony and healing any problems. Take a moment and let your imagination develop a clear feeling for this happening in any way that makes sense to you ... ...

Now whenever you are ready to return to a state of full waking alertness, simply take two deep breaths. Gradually open your eyes, and stretch if you have a need. Don't forget to congratulate yourself for participating in your health and well-being by relaxing and visualizing your body's defenses.

HYPNOSIS AND SELF-HYPNOSIS

People have countless misconceptions about hypnosis. Many people are skeptical or fear it. They feel someone else will control their mind. Others feel it is a magical bullet that will enable them to make changes without any effort on their part. The fact remains that hypnosis can be useful in a variety of medical situations, especially in cancer. I have used it to help people quit smoking and reduce their experience of pain.

An example of the power of self-hypnosis is illustrated by a young breast cancer patient who I saw a number of years ago. When this thirty-something woman was diagnosed, she also found out she was pregnant. Her doctors were not willing to start her chemotherapy until after her first trimester and then they were unwilling to give her any medication to curb her nausea. I was able to work with her for six weeks and taught her how to use a taped exercise to put herself into a trance state. I worked with the nursing staff to coordinate her chemo treatments so that she could lie down and start her tape for about ten minutes before her drugs were injected. She did not have a single treatment or post-treatment episode of vomiting or nausea and had a perfectly healthy baby. She used the tape for that, too!

Hypnosis is known as a trance state. This state is different than our normal waking consciousness, but it is not something weird or abnormal. We often go into trance states intentionally or sometimes by accident. When we daydream about a peaceful place we have been or fantasize about a future event, we may go into a trance state. Whenever we get deeply absorbed in reading a good book or involved in a movie and are not aware of the surroundings, we are actually in a trance state. A third example may happen when we are doing or experiencing something repetitive, like the yellow lines on the highway.

Although some clinicians might disagree, I view relaxation, visualization (guided imagery), meditation, self-hypnosis, and hypnosis as basically the same. They are variations on structures to make it easier for a person to enter a state of low arousal and healing. All these techniques or structures help reduce "static" and critical/analytical cognitive faculties, so that we become more open to suggestion and re-programming.

It is important to clarify certain misconceptions that people have about hypnosis and self-hypnosis.

Common Misconceptions

➤ **Misconception:** *"Hypnosis is sleep."*
Truth: People do not faint or go unconscious. Most people report awareness of what is being suggested and report a state of significant relaxation. This is often verified by physiological monitoring.

➤ **Misconception:** *"You will not wake up from hypnosis or self-hypnosis."*
Truth: You do not go to sleep when hypnotized and may remain aware of certain aspects of the environment. People come out of trance states naturally by themselves if they are left alone. A trained hypnotist will also lead the person back to a state of full alertness.

➤ **Misconception:** *"Only weak-willed, gullible, or unintelligent people are able to be hypnotized."*
Truth: Actually, people who have vivid imaginations, are creative, and intelligent are most likely to be hypnotized.

➤ **Misconception:** *"Hypnosis will cause you to lose control."*
Truth: You will not do anything under hypnosis that you would not do when in full waking alertness. People with control issues or who have issues with trust may not allow themselves to be hypnotized. Actually, hypnosis and self-hypnosis help you gain greater control by incorporating suggestions to help quit smoking or change other habits.

There are basically three steps to hypnosis or self-hypnosis. The first is the **hypnotic induction**, which leads the person to a state of profound, pleasant relaxation. This is established by a structure that enables you to refocus inwardly instead of being bombarded by your environment. Most commonly, I use the eye-fixation technique where a person gently focuses on a spot to anchor his or her attention. By focusing on the spot, your mind can begin to clear without trying and when your eyes are tired, they will close and open you up to healing suggestions. This pleasant state of relaxation can be increased by slowing down the breathing or using imagery like floating on a cloud, becoming heavy enough to sink into the chair or bed, or slowly descending a stairway.

The second stage is the **hypnotic suggestions**. Success is more likely if you word suggestions positively and use imagery like "I will begin to feel a pleasant cool numbness where my pain is." Make suggestions flexible and non-judgmental. Avoid "shoulds" and "musts" so you don't rebel or feel additional pressure. Example: "I will protect my body by controlling my desire to smoke." Make suggestions gradual rather than immediate and repeat the suggestions. Avoid implying doubt or failure. Examples:

- "As my breathing slows down, everything begins...to...slow...down."
- "With each breath, I become more and more at peace, more and more relaxed, more and more calm."
- "I can see my inflamed red tissue gradually becoming pink and healthy."

The third stage is **ending the hypnotic state**. It is best to slowly return to full waking alertness. I usually have people count from three to one. *Three, take two deep breaths, and one – slowly open your eyes whenever you are ready. Stretch if you need. Continue to feel relaxed and renewed.* As mentioned before, even if you don't do anything in particular, you will return to a state of full waking alertness.

Here are several types of images that have been successful in my work. The first I call **Magic Ingredient**. This is where a person imagines a powerful effect from a normal experience. I had been working with a highly suggestible person with lymphoma who had severe stomach pain. I noticed that she often drank large glasses of water particularly before and after her treatments. I suggested that she imagine she could reduce her pain anytime by drinking "magic water" that would form a cool numbing coating from her throat to her stomach. A variation was used with another person who had phlebitis as a result of her chemotherapy. As part of her treatment, she began wearing special support stockings and was instructed to imagine these "magic stockings" reducing her pain as she slowly pulled the stockings up her legs.

Another common image is **Glove Analgesia**. After a person is guided through relaxing imagery, he/she is asked to imagine sitting at a table, which has a clear glass bowl containing an extremely powerful anesthesia. The person is then asked to imagine placing his/her fingertips into the solution and feeling a cool numbing sensation, which can spread up throughout the body or be transferred by a touch of that hand. This is the basis of the transcript that is supplied later in this manual.

A third technique is what I call **Image-Counter Image**. The person is asked to give his/her pain a concrete form based on its qualities. Once a vivid image is formed, the person is then asked to imagine it gradually being converted to a less painful and healthy state, An example would be: imagining the pain as a fierce little creature with its teeth or claws digging into that area of the body and then gradually loosening its grip.

A variation on the above technique is what I call **Opening**. In this strategy, the person is asked to imagine an opening like a mouth over the area of pain. The person "breathes cool, numbing breath in and pain, discomfort, and toxins out. A similar technique is called **Projection**. This is where the person imagines the pain in the form of a large colored ball (usually red) and projects it out of the body where it can be shrunk and gradually converted to a cooler color (usually blue) and then brought back into the body under better control. The Simontons describe this technique in more detail in their book, *Getting Well Again*.

One last technique is **Subject-Object**. In this strategy, the person disassociates (e.g. mentally floats up to the ceiling) and views the pain or discomfort as happening to someone else. This "out of body" experience is especially useful with uncomfortable or painful procedures like chemotherapy injections or bone marrow aspirations.

There are many more types of images and variations that can be used. You can mentally rehearse these or make tapes personalized to your situation. As mentioned earlier a transcript that you can use as a model is on the following pages. You may substitute warmth instead of coolness, if you desire. A simple self-hypnosis technique is also supplied.

A SELF-HYPNOSIS EXERCISE FOR RELAXATION AND PAIN CONTROL

Begin by assuming a comfortable position. You may be sitting down or lying down. Take a moment to adjust yourself, so that you are feeling comfortable and at ease. To begin, gently pick a spot in your line of vision, and focus on that spot. As you begin to focus on that spot, take a moderately deep breath and hold it for a moment. Breathe in gently, smoothly, and quietly. Now, let loose as you exhale fully. Again, as you stare at the spot, take in a moderately deep breath and hold it briefly. Again, as you let the breath out, feel your body become less tense, as you expel the air from your lungs. Now, take a third moderately deep breath and hold it momentarily. Again, let the tension out with the breath as you exhale fully.

Now, allow your breathing to gradually slow down. Notice, as the tempo of your breathing slows, everything slows down. Becoming more and more at ease. More and more at peace. Slowing down. Everything is continuing to slow down. As you focus on the spot, focus your attention on your hands. You may notice that your hands are feeling pleasantly comfortable and relaxed. As you continue to stare at the spot, notice that your breathing is slowing down. As your breathing slows down, notice that you continue to slow down and feel more and more at peace, more and more at ease, more and more calm. Notice that as you continue to stare at the spot, that your hands feel more relaxed and comfortable. Focusing on the spot and the sound of my voice, if your eyes become tired or heavy, just allow them to close and let your breath continue to become slow and even, steady and regular. Despite any distractions that may be around you, continue to focus on the spot or close your eyes and focus on the center of your forehead as you continue to slow down. Everything is slowing down.

In just a moment, I will begin counting backwards from five to one. As I count notice that you can deepen your state of relaxation. With every breath you take, notice that you become more and more calm and quiet. With every breath out, you exhale tension and worry.

- **Five**, feeling more and more at ease, more and more at peace, quiet, calm. More and more relaxed, more and more quiet. More and more calm. If you have not closed your eyes, please let them gently close now so that you may experience a deeper state of peace and be less distracted.

- **Four**, slowing down, more and more relaxed, more and more at peace, more and more at ease. As your breathing slows down, everything is ... slowing ... down. Notice your hands feel quiet and comfortable.

- **Three**, more and more relaxed, more and more at peace, more quiet and calm. Breathing in more and more relaxation, more and more peace. Slowing down, everything ... is ... slowing down.

- **Two**, continuing to slow down. Everything ... is ... slowing ... down. Quiet, peaceful, at ease, relaxed, calm. More and more relaxed, more and more at peace. Your hands feel comfortable and relaxed.

- **One**, slowing down. Feeling more relaxed and peaceful. Everything ... is ... slowing ...down. Quiet, calm. Becoming more peaceful and relaxed than you ever thought possible. Ready to open up to new and interesting experiences. Feeling peaceful, quiet, calm, and at ease. Opening yourself up to new and interesting experiences that you can picture in your mind's eye.

And now, to begin, I want you to notice how deeply relaxed you are. Become aware that you have the power to decrease your pain. Now imagine that you are sitting at a table. In front of you is a clear glass bowl. In it is a bluish liquid that is a powerful anesthesia. As you stare at this liquid a light shines on it and gives it a bit of a glow and you know that this can reduce your pain. Now you gently reach out your hand and lightly touch it to this bluish liquid. Notice as your hand touches this liquid that it becomes pleasantly cool and numb. Feel this liquid surround your hand making it cool and numb in a pleasant way and feel this sensation begin to flow up your hand. Feel this pleasant coolness flow up your hand and into your arm. Feel it flow up your arm and into your shoulder. Feel this cool, numbing sensation flow across your shoulder and down your other arm and into your other hand. Feel this cool, numbing sensation flow into your neck and up into your head. Also feel it begin to slowly flow down your neck and into your chest making it light and easy to breathe. Feel it flow slowly down your back soothing and numbing each vertebrae. Now, take a moment to notice that your whole upper body is pleasantly cool and numb...

Now, notice that this cool pleasant numbness begins to slowly flow into the lower half of your body. First, around your lower abdomen and hips and into your thighs and your knees. Feel them become cool and comfortable. Now, feel this pleasant cool sensation flow into your calves and finally your feet. Feel this cool, pleasant numbness flow into any areas of your body that still feel pain or discomfort until that eases. Now take a few minutes to stay suspended in this state of pleasant, cool numbness for as long as you like...

Whenever you wish to return to a state of full waking alertness, count from three to one. **Three**, take **two** deep breaths. Begin to open your eyes, and **one**, feel refreshed and renewed. Gently stretch if you need to. Congratulate yourself for participating in your health and well-being, by attempting to control your pain.

A SELF-HYPNOSIS TECHNIQUE

There are many ways to induce a state of self-hypnosis. This is a technique developed by Dr. Roger Bernhardt, co-author of the 1977 book, *Self-Mastery Through Self-Hypnosis*.

This technique may take as little as 30 to 60 seconds, after some practice, to complete. Learning to allow your muscles to relax is a preliminary step. You can then follow these steps and repeat them as often as possible to enhance the effects.

❶ Pick a specific goal (i.e., a behavior you would like to change, such as stop smoking or eat more).

❷ Follow the induction process:

    a.   Allow your eyes to gaze up toward your eyebrows.

    b.   Slowly close your eyelids and take a deep breath.

    c.   Exhale and let your body float.

❸ Focus on a pleasant scene.

❹ Now repeat a three-part suggestion to yourself, as in this example:

    a.   For my body, not eating (or smoking) is ruinous.

    b.   I need my body to live.

    c.   To the degree I want to live, I will preserve and protect my body.

❺ Picture yourself as you would like to be: mastering situations that you felt previously unable to handle.

❻ Make a post-hypnotic suggestion to help you carry over your commitment to change.

❼ Allow yourself to come out of your trance by counting from three to one.

        Three: Take two deep breaths.

        Two: Gradually open your eyes.

        One: Stretch gently.

HUMOR

"Cancer is probably the most unfunny thing in the world, but I'm a comedienne, and even cancer couldn't stop me from seeing humor in what I went through."
–*Gilda Radner*

I remember seeing an ad with the attention grabbing title "YOU WOULDN'T LAUGH AT CANCER, WOULD YOU?" The ad was directing people not to make fun of mental illness. While you should not minimize or discount the seriousness of cancer, it is vitally important to see the humor that can emerge from your experiences with it, just as humor can be found in life's most profound experiences—love, parenting, learning. Laughter and humor can benefit our health in many ways, whether we have cancer or not.

Folklore has always told us that "He (she) who laughs, lasts." (Mary Pettibone Poole) It wasn't until Norman Cousins publicized the benefits of laughter in his book, *Anatomy of an Illness*, that researchers began to systematically study its effects. He recounts his experiences with a life-threatening disorder called Ankylosing Spondylitis. Cousins related how laughing at Marx Brothers movies and tapes of *Candid Camera* gave him hours of pain relief. Laughter has been shown to increase our natural painkillers. There are other benefits from "internal jogging" such as increasing circulation and exercising lungs and muscles. William F. Fry, M.D., an emeritus clinical professor of psychiatry at Stanford University Medical School, found that laughing one hundred times is roughly equal to a regimen of ten to fifteen minutes on exercise equipment.

Laughter has been shown to reduce stress hormones and decrease tension. A forty-two-year old Hodgkin's lymphoma patient developed an elaborate way to laugh at her cancer and ease her children's fears. Every time she would go for her chemotherapy treatments, she would place aluminum foil over her bald head and attach some antennae to pretend she was from the planet, Materamus. She had a great time interacting with her kids and it really helped ease the effects of her treatments.

Humor and laughter helps us face difficult times. They are good weapons to counter fear and trauma and can help restore perspective and generate hope. Even in the most dire of circumstances people find humor. There were jokes told in the concentration camps. The film, *It's a Beautiful Life*, shows how a father's humor helped protect his son from the horrible truths of those camps. Erma Bombeck wrote humorously about childhood cancer in her book, *I Want to Grow Hair, I Want to Grow Up, I Want to Go to Boise*. She was also able to laugh at her own cancer:

> "... a local anchorwoman was touting a special on women
> with breast cancer and said in the teaser, "There are ways

to help you become a whole woman after mastectomy."
The words "**whole woman**" made me crazy... The part of me
that is missing didn't think, laugh, or contribute a single
thing to this planet. I still form coherent sentences and
intelligent opinions. I defy anyone to read my columns and
tell me which ones were written by a single-breasted writer."

Humor also helps people release anger and increase creativity. Woody Allen developed some zinging one-liners out of his neurotic fears. Some examples:

It's not that I'm afraid to die, I just don't want to be there when it happens.

It is impossible to experience one's own death objectively and still carry a tune.

Eternal nothingness is okay if you're dressed for it.

The universe is merely a fleeting idea in God's mind—a pretty uncomfortable thought, particularly if you've just made a down payment on a house.

There are some things that are worse than death. If you've ever spent an evening with an insurance salesman, you know exactly what I mean.

Lately, it has been found that humor and laughter benefit our immune system. Herbert M. Lefcourt, Ph.D., a psychology professor at the University of Waterloo in Ontario, Canada found that humor increased salivary immunoglobulin A, which is a first line of defense against viral and bacterial infections. Obviously, restoring a state of health is more complex than *Take two jokes and call me in the morning*." However, developing a sense of humor may help you keep your balance.

As Wavy Gravy, that Merry Prankster from the '60s said, "It only hurts when you don't laugh. Laughter is like the valve on the pressure cooker of life. You either laugh at stuff or you end up with your brains on the ceiling. If you don't have a sense of humor, it just isn't funny anymore. I forget who said that stuff, but I'll testify to its authenticity."

DEVELOPING A SENSE OF HUMOR AND CREATING MORE OPPORTUNITIES FOR LAUGHTER

Two helpful resources for this topic are *The Laughter Prescription* by Dr. Laurence J. Peter, and *Heart Humor & Healing* edited by Patty Wooten, R.N.

❖ Remember to develop an attitude of playfulness. Spend time in toy stores.

❖ Think funny. See the flip side or extremity of outrageous thoughts.

❖ Laugh at incongruities.

❖ Only laugh at others for what they do, rather than what they are.

❖ Jokes need to be playful, not bitter. It is more important to have fun than to be funny.

❖ Laugh at yourself and others with objectivity and acceptance, not derision.

❖ Take yourself lightly. Take your job and your responsibilities seriously, but have fun.

❖ Make others laugh and encourage them to joke as well.

❖ Cut out and share cartoons or humorous sayings.

❖ Watch funny movies, listen to stand-up comics, or read the comics.

❖ Keep playful things around like funny glasses or wind-up toys or bubbles.

❖ Make a list of your favorite comedians, movies, and things that make you smile and refer to it from time to time.

❖ Recall since your diagnosis if anything funny (maybe not at the time!) happened. Write it down.

SOME ADDITIONAL STRESS MANAGEMENT SUGGESTIONS

There are many other stress management suggestions and hints people with or without cancer use. This manual cannot hope to cover them all. This section will survey some other ideas that might prove useful.

Obviously, cleansing the body and promoting good general physical health can help you manage stress. Moderate physical exercise helps strengthen your body and relieve tension. It also may have immune enhancing effects. Researchers have found that cancer patients during chemotherapy and radiation treatments may reduce side effects like nausea and vomiting or ease pain when they engage in modest exercise. This may include walking or gentle stretching. Movement such as dancing may be liberating and help let go of tension. One research study found yoga increased immune function in breast cancer patients. It is beyond the scope of this manual to prescribe an exercise regimen. Exercise programs need to be developed for each individual and certain warnings do need to be considered especially when a person has metastatic bone disease or is taking a chemotherapy that is cardiotoxic (causes heart damage). Always check with your medical team before beginning a program and start gently.

All our sensory modalities can be accessed for healing and stress relief. Touch is very healing. Many cancer patients are "touch hungry." Gentle touch like **slow stroking** (repetitive light moving touch over a body area) can relieve pain or ease anxiety. A compassionate touch like holding hands or a gentle hug can show support where words fail. Let people know if you want to be touched. Massage is a great way to relieve muscle tension and let go of stress. There has been some research that has also shown massage to enhance immune function. There are several different types of massage and many certified and legitimate massage therapists. You can also learn to do self-massage. **Acupressure (acupuncture** without needles) and **shiatsu** (a type of finger–tip pressure point practice) may be helpful avenues to relieve tension and pain and restore energy flow. There are other types of bodywork practices that may be beneficial. **Reflexology** is a type of foot massage that helps detect and correct energy imbalances. **Therapeutic touch** is another modality, which has received study for its healing properties. Actually, therapeutic touch involves the energy fields of the body and movements are made over body areas without literally touching them.

Different sounds, vibrations, and music can have healing and stress-reducing effects. Dr. Mitchell Gaynor does a good job of examining this in his book, *Sounds of Healing*. Chanting repetitive tones or prayers certainly can have a very soothing effect on us. Certain tones that vibrate through our bodies and minds can also be calming or stirring. Music, too, can have powerful and beneficial effects. There is a great deal of engineered music that can guide us to deeper levels of relaxation. Some slow classical pieces can generate the same effect. Some studies done by music therapist and clinical

111

psychologist, Mark Rider, Ph.D., have found increases in antibodies and different groups of white blood cells. Some people enjoy nature sounds—either "live" or recorded.

Our sense of smell is our most primitive and emotional sense. Certain smells excite us. The perfume industry spends millions to find the scents that do. Certain smells may even make us sick. There are many examples of cancer patients developing conditioned nausea from smells that they have associated with the clinic where they are treated. An example is an alcohol smell or a nurse's perfume. This can still happen years later. Now we are finding that certain aromas can be relaxing. Through brain wave studies several scents have been found to have a quieting effect. Apple pie is one of the most relaxing. The whole area of aromatherapy has been revitalized with such findings. Lavender is a particularly relaxing smell. Eastern meditators use incense to help them ease into a relaxed state. There are many ways to produce an aroma, such as potpourri, incense, scented candles, air fresheners, and essential oils used to help create a more relaxing environment.

Our sense of sight has come to dominate our senses. Filling your environment with pleasing visual stimulation can also help create a sense of calmness. That may involve reducing clutter and simplifying things. It may mean putting up attractive posters or art. It may mean a great view out your back yard window. Arranging your environment to include pleasing visuals is another helpful way to add relaxation. This does not have to be expensive.

There are other complementary medicine approaches to stress management such as herbs and vitamin supplements. Kava Kava (recent research has found that Kava Kava may cause liver problems), Valerian Root, St. John's Wort, SAMe, the different ginsengs, and antioxidant combinations are being touted as helpful to relieve both mental and physical stress. More research is being done and looks promising. However, "natural" does not necessarily mean safe. Please, discuss what you are doing with your medical team to make sure that nothing might interfere with your treatment plans.

"LIVING BEYOND LIMITS"

Does Social Support Lead to Longer Life?
And, How Much Does that Matter?

David Spiegel and his associates at Stanford reported a landmark study with metastatic breast cancer patients. Women in the support group arm of the study lived significantly longer than those in the control group. These results made a believer out of Spiegel who ironically doubted the influence of psychosocial factors on physical health before his research. He later wrote a book on his clinical experiences entitled, *Living Beyond Limits*.

These findings remain controversial and have triggered a debate which continues to rage as to whether social support for cancer patients leads to increased survival. One thing appears certain: social support improves psychological adjustment to cancer. It also appears to reduce pain. Since it improves the quality of life a great deal, does it matter whether it also extends life? Can we afford not to tap into its positive effects?

A cancer diagnosis is one of the most traumatic experiences that can happen to someone. You shouldn't feel the need "to go it alone." Within reason, you can never have enough support and support comes in many forms. Sometimes, too much "hovering" can feel like smothering. In this section of the book, we will focus on two major sources of support – the social network, particularly family and friends, and support groups.

The support these very important individuals and groups offer falls into three principle categories: material or physical, informational, and emotional or psychological. Each of these is necessary and may come in a variety of forms. Members of your social network may be better at one particular form of support or may be good at all three. It's important to be sure that your expectations are not too unrealistic.

Examples of material support are financial aid, transportation, meals, physical treatment and medical supplies.

There are two main types of informational support, **reference information** and **inspirational information.** Reference information is factual information that is accurate according to the textbooks and the clinical experience of medical professionals. This is rational information usually gleaned from large groups of people, otherwise known as statistics. This would be information obtained from the National Cancer Institute or the American Cancer Society. If we trust our health care providers — and we should — then we need to welcome their wisdom, but not necessarily follow it blindly. We also have our own inner wisdom. We usually refer to this as intuition or intuitive wisdom. We make the best decisions when we put these two realms together.

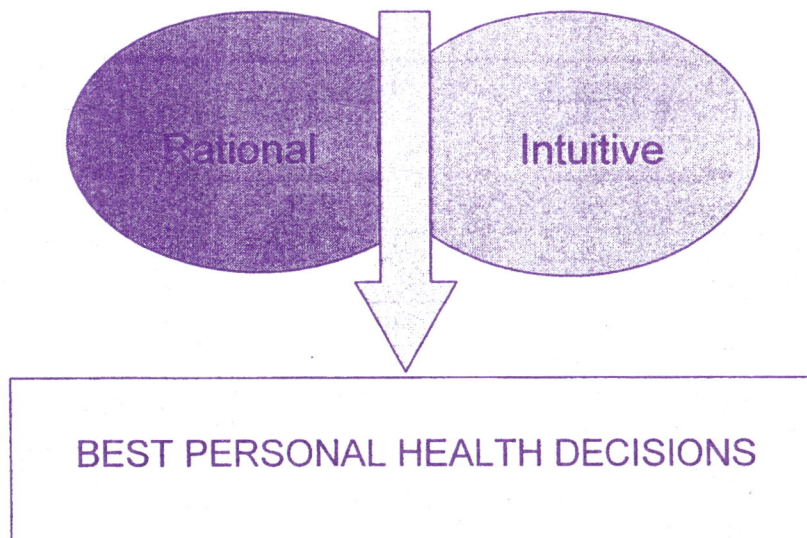

Rational   Intuitive

BEST PERSONAL HEALTH DECISIONS

Inspirational information speaks more to us on an emotional level. Lance Armstrong's story *(It's Not about the Bike)* is a great example. It is not that many of us could even qualify for one Tour 'de France or get over our cancer if it were localized. It is the hope that if one person could beat cancer, then I can do it! Some other examples would be: Bernie Siegel's books, *Love, Medicine, and Miracles; Peace, Love, and Healing,* and *How to Live Between Office Visits; Chicken Soup for the Surviving Soul* by Jack Canfield, Mark Victor Hansen, Patty Aubery and Nancy Mitchell, R.N.; *There's No Place Like Hope* by Vickie Girard; and *From this Moment ON* by Alice Cotter. Some resources are both inspirational on an emotional level and substantive in stretching the body of what we know rationally. Hamilton Jordan, author of *No Such Thing as a Bad Day,* survived three different types of cancer. He shares his experiences, but also offers some helpful suggestions that are factual and informative.

There is so much information now available that it is possible to become too saturated. You can't read everything and you can't follow up every doctor, clinic, or supplement people suggest. If you get overloaded, stop reading for a while. Come back to it when you feel ready. That is where a practice of daily quiet can help you "intuit" what is best to pursue at any given time.

The third type of support is emotional or psychological support. This can be the hardest for people, because they often don't know what to say or do around cancer patients. The patient may need to do some educating as to how he or she needs or wants to be treated. You may also need to spend more time with people who are positive and restrict your time with those who make you uncomfortable or "bring you down."

Identify your sources of support for the different categories.

Material Support	1)_____ 2)_____ 3)_____	1)_____ 2)_____ 3)_____	1)_____ 2)_____ 3)_____
Informational Support	1)_____ 2)_____ 3)_____	1)_____ 2)_____ 3)_____	1)_____ 2)_____ 3)_____
Emotional Support	1)_____ 2)_____ 3)_____	1)_____ 2)_____ 3)_____	1)_____ 2)_____ 3)_____

Cancer is a social disease and even though one person contracts it physically, the whole social network is affected. Although it can certainly have a significant impact on co-workers and friends, the family is most deeply affected. While there may be some common shared responses, each and every family member is likely to have their own unique set of feelings, and will likely also express those feelings differently. This is often based on what their past experience with cancer has been or on certain beliefs they may have about illness. Entire books have been written about family reactions to cancer. This section will try to touch on the most important points.

Roles in the Family

Everyone in each family has certain roles or functions within the family. For example: the mother is often the **nurturer** or the main support to "kiss it and make it better." Other examples are the joker, the breadwinner, the fixer, the black sheep (or independent one), the leader, the dependent one, the scapegoat, etc. Cancer can certainly affect family members' role by changing them or reinforcing them.

What was your role or roles before the diagnosis of cancer?

_____

_____

_____

How has your role changed?

_____

_____

_____

What are the roles of the other family members? How have they changed? And, how have they dealt with, coped with, or been affected by the change? Use the following chart to reflect on these important family questions.

Family Member	Role	How changed? How has he or she done?

Family Reactions

As mentioned earlier, each family member may experience different feelings and choose to express them differently. There is such a wide range of feelings that it is beyond the scope of this book to explore. What is most crucial is not to read minds or assume you know what others are thinking or feeling. Instead, be sure to check out a belief you may have directly. Start by clarifying your own feelings. This is a three-step process.

1. Label your feelings.

This is a pivotal step, whether it is strictly internal or interacting with others. Labeling accomplishes two functions. First, it clarifies what you are feeling and may remove doubt. Not everyone who gets angry yells or throws things. Sometimes anger is mistaken for sadness. Labeling the feeling removes confusion. The other function labeling serves is to convey the feeling without charging it with emotion. For example:

"I get irritated when my 'good' friends don't visit."
"I get angry when my 'good' friends don't visit."
"I get furious when my 'good' friends don't visit."

Notice that the intensity of the feeling is conveyed with the choice of label. When statements are emotionally charged with yelling or screaming, people focus on the emotion instead of the issue.

2. Attach the feelings to specific sources.

While it is important to state a feeling, it is unlikely to resolve anything, unless you can identify something concrete enough to change. By identifying a specific source, you then can work on problem solving or acceptance of a situation. This is shown in the example by attaching the feeling, anger, to "good" friends who don't visit.

3. Identify what you need to do to express the feelings or to accept the feelings.

The third step is crucial for you to discharge the feeling in some constructive way. In the example above, the person can phone a friend and make a date to meet. If the friend does not want to come visit, then it may be necessary to let go of that relationship or at least your expectations about support at this time.

Communicating feelings, needs, and expectations is most crucial in the family. Family members' reactions and expressions may not be what you thought they would be. You may have to educate people even if you don't want that job. You will be treated the way you allow yourself to be treated, unless you speak up.

How have family members, friends, or co-workers reacted? Who has pleasantly surprised you? Who has disappointed you? What would you like to do to make positive changes in your relationships?

Family Member, Friend, Co-Worker, etc.	Constructive Reaction	Unhelpful Reaction	What positive changes?

Make a list of supportive things people have said or done. Make another list of things people have said or done which you didn't feel were helpful. Decide how you need or want to respond to each of these.

Helpful/Supportive	Unhelpful/Nonsupportive

What is the best advice would you give another cancer patient about dealing with family and friends?

_____

_____

_____

_____

_____

_____

_____

Closeness and Disclosure

You are closer to some people than others. You might not feel comfortable giving out all the same information about yourself to everyone. To whom are you closest? To whom are you more distant?

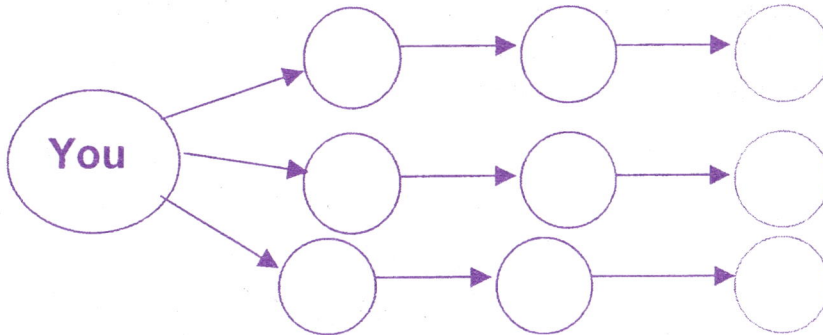

You may want to write initials in the circles to represent different levels of closeness.

When we give out personal information, we refer to that as **disclosure**. You have a right to decide whom you want to know what about you (Does this sound like Abott and Costello's famous comedy routine, "Who's on First?"). You may not want to disclose a great deal of information to certain people. You may get tired or anxious to have to repeat the same information time after time. Some families appoint a "press secretary" to keep everyone informed. You are not rude if you choose not to talk about your cancer to certain people and it may not mean you are in denial. It is actually healthier to talk about other areas of interest and to keep your relationship normal or work on growing closer. There is more on this in the chapter on assertiveness.

The Family Meeting

One of the best suggestions I can offer to a people is **The Family Meeting**. After some of the initial shock has worn off, gather as many of the family members as possible at a central location. Then lay all the known information on the table and allow people to ask questions and discuss their feelings. Next is the most important task and that is to reassign roles and parcel out responsibilities as to how the family hopes to deal with this cancer.

Having a family plan that delegates and spreads duties among willing family members (and friends) can cushion against "burnout" or undue stress on any one family member. These duties or responsibilities should be parceled out according to people's strengths rather than people being "guilted" into something. For example: Some people may have a really difficult time going to the hospital or the cancer clinic. It may be better to have them do laundry or make meals instead of "forcing" them to take the person for chemotherapy.

Everyone in the family needs to remember that this is a marathon rather than a sprint. It may be useful to set up a schedule for regular meetings (at least for the immediate family) even if cancer is not a topic on the agenda. Cancer doesn't end just because treatment does. Once cancer is diagnosed, it becomes part of one's life. How big a part is up to the person and the family. You can use the chart below to assign responsibilities.

Family Member/ Friend	Task/ Responsibility

Support Groups

There are many sources of emotional support besides family. There are friends, neighbors, co-workers, and religious affiliation. One very important source of support for cancer patients can be a constructively run cancer patient and/or family support group.

Unfortunately, a lot of people who could benefit from support groups avoid them. People feel (rightfully so) that their family or place of worship offers them all the support they need. Cancer patients often have misconceptions or fears about going to a group. They expect to see sick people who are wasting away and complaining about all sorts of things. They often are surprised after attending a good group. Yes, intense and tough issues that may provoke intense or tearful reactions are discussed, but people in the group also laugh and discuss the good things in their lives such as their hobbies and grandchildren.

There was a wonderful cartoon in *The New Yorker* magazine (1/18/93) that showed a bearded, dehydrated man under a blistering desert sun crawling on his hands and knees past a group of people. The leader proclaims, " Sorry no water. We're just a support group." If support groups don't offer cancer treatment such as radiation or chemotherapy or extend life, exactly what do they offer?

As mentioned earlier research studies have had mixed results about extended survival related to social support or attendance at support groups. There are several studies that show increased survival or improved immune function. There are other studies that report no difference. Research is more consistent when it comes to improved adjust to cancer diagnosis, less anxiety and depression, and decreased pain.

Support groups have several unique features that family members and others can't offer. The most important feature is that people in the group really do understand what the person is going through because they are going (or have gone) through it, too. A second factor is that other survivors (especially long term ones) inspire the patient more than all the positive statistics or prognostic pronouncements medical staff can utter. A third uniqueness is it provides a safe and confidential environment for a patient to discuss delicate issues without judgment and less emotional attachment. An example might be where a patient wants to talk about death and the family members may interpret that as giving up or being negative because they are afraid. Their response might be "Oh, you won't die from this!" which sends the message not to talk about that. A support group might offer a chance to identify fears or resolve some issues that the patient can then take back to the family.

For these reasons, I encourage people to explore local support groups for cancer patients. While there are good patient run groups, it is often more constructive to have a professional run the group or at least attend or consult at times. There are groups for specific types of cancer or groups for all types of cancer mixed together. Some are very structured and some are very loose and discuss issues that are most immediately on people's minds. There are also support organizations like **Gilda's Clubs®** or **Wellness Communities®** that offer programs. Be sure to check resources in your local community.

Another source of support is one-on-one contact, where a cancer patient that is through with treatment and in remission might meet with someone who has that type of cancer and is newly diagnosed. This type of support may be an alternative for people who really dislike groups. Again, check in your local community or cancer clinic for such programs around you.

EXPRESSION OF FEELINGS:
SOME THOUGHTS ON ASSERTIVENESS

Keeping your feelings under tight control takes a great deal of energy. It takes even more energy to hide your feelings behind a smiling mask, especially if you don't feel that way. Moreover that energy is wasted, like heat going up your chimney instead of warming your home. The same is true when you don't honestly deal with your feelings and try to hide them. It is wasted energy that can be redirected into healing. An important way to do that is to be assertive, express your feelings, and stand up for yourself. Assertiveness acts like the valve on a pressure cooker, so feelings do not build up and become explosive, unproductive, and out of proportion.

While a course in assertiveness is beyond the scope of this manual, it is important to become appropriately assertive in your social relationships so that you don't waste your energy. Assertiveness is often confused with aggressiveness, but the two are very different. The focus when being assertive is on communicating your feelings and expressing that they are at least as important or valid as others. Being assertive does not attack anyone and is not overly emotional. If the focus of assertiveness remains on "sticking up for yourself" instead of changing others, you will feel good about yourself and if they change, it will be an added bonus. If you focus on changing others, you may end up more frustrated, if no change occurs. Remember that the focus is on gaining respect for yourself, so that people will not mistreat you. Much of the time people don't know that, unless specific feedback is given, they cannot know what behavior to change.

An example might serve to illustrate: Jane (not her real name), a forty-year-old breast cancer patient, was angry and hurt that some close friends of her and her husband had stopped contacting them. She brought this up in therapy and at first did not want to call them. She felt that "she was the sick one and that they should call." We discussed that while some of that feeling might be true, not calling them was self-defeating. She agreed. At her next appointment, Jane happily reported that she had followed through. She had called and when her friend answered, she said, "Hi, I'm not dead yet! Let's get together and catch up." This had humorously broken the ice and they had dinner and caught up on what had been happening. While they did talk some about Jane's disease, the focus was on missing their friendship. Had she not been assertive, Jane might still be waiting for her friend to make the first call.

You can learn more about assertiveness by reading classic books such as *Stand Up, Speak Out, Talk Back* by R.E. Alberti and M.L. Emmons (1975) or *Don't Say Yes When You Want to Say No* by H. Fensterheim and J. Baer (1975).

Assertiveness with your medical team is especially important. Nowhere is it more critical to communicate effectively. Here is an example where being unassertive caused needless stress. A seventy-two-year-old breast cancer patient started her treatment and seemed to be coping generally well. She was quite pleasant to staff and brought cookies to the nurses who amiably fought over who would treat her that day. After four or five treatments, Helen (not her real name) became increasingly irritable, would only let certain nurses treat her and got so upset before her treatments that she vomited in the parking lot. The nurses asked me to talk with her and in the course of our one and only interview, she said, *"I don't understand why I have to get this nasty chemotherapy. My doctor said my cancer was gone."* Unfortunately, she had not discussed this with her doctor and had become quite resentful. We did a short role-play and she got up enough nerve to question him. He was more than happy to sit down and explain that women with her type of breast cancer had less chance of recurrence if they received the chemotherapy. This made sense to her and she reverted to her pleasant self.

Talk to your medical team. It is important to be partners in this process of cancer treatment. You need to feel comfortable and confident with them. Let them know what is going on for you rather than waiting for them to ask the right question. Teach them how you need to be treated and let them know when they haven't treated you well. We are all human and assume we are behaving in an okay fashion unless we receive feedback to the contrary.

Assertiveness is just one way to express and discharge feelings constructively. Assertiveness is very situation specific. Sometimes we need to process feelings before we immediately assert ourselves. We may process things and feel we don't need to say or do anything and can let it go. This needs to be a conscious decision rather than suppressing our feelings out of fear of being disliked. The main rule of thumb that I use with patients is this: If you keep rehashing a situation and find yourself generating intense feelings, then you probably need to assert yourself and get some resolution. If you can let the situation go without rethinking about it or reminding yourself to act differently next time, then you probably don't need to assert yourself. The next two pages summarize some considerations when being assertive.

SOME IMPORTANT CONSIDERATIONS WHEN BEING ASSERTIVE

❶ The intention behind assertive behavior is to stand up for yourself and express your feelings, not to hurt the other person. Keep in mind that sometimes other people may get hurt no matter how diplomatic you are. Don't expect other people to know how you feel. Other people feel their behavior is okay unless they hear otherwise from you. You will be treated the way you allow yourself to be treated. Examples of assertive responses:

> ➤ "I can only handle a 20 minute visit."
> ➤ "I'd rather visit about other things and not just talk about cancer."
> ➤ "I really need to talk some about my disease."

➤ "It's not helpful when you tell me about other people who have died from cancer."

❷ Keep the focus on "I" statements rather than blaming others. For example: "I get angry when you don't let me do what I can for myself," rather than "You make me angry." The latter tends to make people immediately defensive and not hear your message. The former is specific feedback, so you can clarify what you can do and what you really need help with.

❸ Keep the focus on communicating what is happening for you or what you need, rather than thinking you are going to change someone. If your focus is on feeling good about yourself because you said what you needed, then you win 100 percent of the time. If your focus is on others changing, then you are likely to get frustrated. If people change, that is a bonus. What is likely to happen if you stand up for yourself consistently is that people will learn to respect you.

❹ Be as immediate as you can, so that feelings don't build up and get blown out of proportion and people remember what you are talking about. When speaking, be as direct, firm, and unemotional (calm) as you can. Getting emotional may cause people to miss your message by focusing on the emotions rather than the issue. It also tends to make them feel defensive, which reduces the likelihood that anything constructive will happen as a result of your efforts.

❺ Be aware of your nonverbal communication. Make direct eye contact. Hold your posture erect but not too stiff. Avoid hesitating in your speech and try not to joke if you are making an honest request. If you are looking away a lot, are slouched, or hem and haw in your speech, people may perceive you as lacking confidence and try to dominate you.

❻ Make your requests direct, concrete and ask for exactly what you want. Recognize that the other person has a right to refuse your request. Be prepared to negotiate. The same is true when someone asks you for something. Take your time to respond, if you are uncomfortable at the moment the request is made. Example: "Let me check when my doctors'appointments are and I'll call you back." Realize that you can offer an alternative and negotiate.

❼ Being assertive is not aggressive. Assertive behavior means that your feelings are equal to the other person's. Being assertive does not mean you are putting your feelings above those of the other person, but it also does not mean they are less important either. Being assertive is respectful of the others' feelings. Aggressiveness does not respect others' feelings and often the intent is to hurt others. Being passive is not respectful to your own feelings. When passive, one gives up control and tends to feel more powerless. It also makes the other person seem more powerful and insensitive. Usually resentment builds and anger can become explosive after a long build-up or can come out in more indirect ways that are usually not constructive. Example: The doctor gives you a prescription for a medication. You don't want it and have no intention of filling the prescription. Instead of talking it out and understanding your doctor's perspective and reaching an agreement, you take the prescription and your doctor thinks you are taking the medication. He or she is puzzled that you are angry when your symptom doesn't diminish.

❽ A brief comparison of assertive, aggressive, passive, and passive-aggressive behaviors:

ASSERTIVE ——————————▶ Emotionally honest, self-enhancing, direct, chooses for self, confident, self-respecting as well as respecting others.

AGGRESSIVE ——————————▶ Emotionally honest, direct, self-enhancing at the expense of others, chooses for others, righteous, superior and derogatory, leaves others hurt, resentful, angry, embarrassed or defensive.

PASSIVE ——————————▶ Emotionally dishonest, indirect, self-denying, allows others to choose, may generate hurt, anger, anxiety or resentment in self, may generate lack of respect from others.

PASSIVE-AGGRESSIVE ——————————▶ Emotionally dishonest, indirect, self-denying, manipulative, chooses for others, may generate anger or defiance in self, anger or irritation in others.

THE CANCER EXPERIENCE

Cancer can certainly produce vivid memories. Some of these experiences you might just as soon forget and others are indelible treasures to hang onto. What do you most vividly recall about your experience with cancer? Below is a chart that can provoke your thoughts back through your journey with cancer.

MOST FEARFUL OR SAD	MOST ANGRY
MOST LOVING	MOST HOPEFUL

JOURNALING
&
OTHER CREATIVE EXPRESSION

Journaling, and other forms of creative expression, including rituals can help us deal with stress and other intense feelings. Journaling is similar to keeping a diary where you write about what is important to you or whatever comes to mind. It is a process that gently invites us to explore ourselves. There are no rules other than those you may set for yourself. Several studies, especially work done by James Pennebaker and his colleagues (1990, *Opening Up*), have shown that journaling about important events not only helps us work through feelings, but can increase our bodies' defenses, and decrease visits to the doctor/health center. Two good resources for journaling are *Writing As A Way of Healing: How Telling Our Stories Transforms Our Lives* by Louise De Salvo (1999) and *The Well-Being Journal: Drawing on Your Inner Power to Heal Yourself* by Lucia Capacchione (1989).

If you like the format, you can easily adopt it and pick meaningful quotations for yourself. Otherwise, you may want to start writing a free form journal. There need not be any rules about how much or how often you must write. In fact, there are no "musts" in journaling. Grammar, punctuation, even complete sentences do not matter. What does matter is your willingness to take time for yourself, in letting thoughts and feelings spill onto the page. Approach it as a helpful process for yourself to explore. Writing poetry may help express feelings and allow their discharge. I found it very therapeutic to write a book of poetry (*Remembering Those Who Sleep Beyond the Dust*) to deal with the loss of my father to brain cancer, my mother to breast cancer after an eight year struggle, and many special people I had meet in my twenty-four years of cancer care.

Writing may not be something you enjoy or are willing to do. Disclosure in therapy or through a support group might be helpful. Try to find local resources. As mentioned elsewhere, some research reports several benefits from therapy and/or support group members such as better adjustment and possibly longer survival.

Other means of creative expression or healing rituals may be a great release. I have many patients draw their cancer, immune system, and treatment, as well as how they see themselves. We discuss these drawings to explore what may be happening subconsciously and to generate positive mental images that they can use in their process of healing. Many clinics have a graduation ritual when a person successfully finishes radiation or chemotherapy treatment. One artist patient who had lymphoma chose to sculpt her tumor and each morning the first thing she did was shave a slight bit of the clay off. This action made her feel like she was fighting back and helped her start her day in a positive way.

Many people like to start out and/or end the day with a meditative or prayer ritual. They find benefit from a gratitude journal. The main goal is that the ritual is meaningful to you and it gives you a source of strength.

On the following pages, you will find some quotations that can spark your thoughts. A **Day-By-Day Monthly Cancer Wellness Journal**© is included as an addendum at the end of the manual.

SOME LINES TO WRITE POEMS OR PROVOKE THOUGHT

⌘   Sometimes the Angel is a Disease. (Remen)

⌘   Unfathomably deep pools of resilience (O'Neil)

⌘   Inside the Miracle (Nepo)

⌘   The sun shines not on us, but in us. (Muir)

⌘   The strange and wonderful are too much with us. (Clampitt)

⌘   Let yourself be silently drawn/by the stronger pull of what you really love. (Rumi)

⌘   There's just no accounting for happiness (Kenyon)

⌘   There's no such thing as a case of Cancer Lite—emotionally. (Edwards)

⌘   I really expected to be dead by now or
I really expected to be alive by now.

⌘   Disfigurement /Need not include/My soul. (Hjelmstad)

⌘   Running so fast that I'm waiting for my soul to catch up. (paraphase Schwartz and Hass)

⌘   ... imagine what you most/would like to do/to help keep the world magical? (Edens)

⌘   All healing is release from fear (paraphrase <u>Course in Miracles)</u>

⌘   Where you have fallen, you stay/ In the whole universe, this is your place./ Just this single spot./ But you have made this yours absolutely. (Pilinszky)

⌘   The whole point is to find your own way. (Campbell)

⌘   The only handicap is a bad attitude. (Hamilton)

⌘   You can't have testimony, without a test. (Vanzant)

Include some of your collected quotations:

⌘   ⌘   ⌘   ⌘   ⌘   ⌘   ⌘

_____

_____

_____

_____

_____

_____

_____

_____

_____

_____

_____

_____

_____

_____

_____

_____

KEEPING CANCER IN ITS PLACE

Cancer has a powerful enough stereotype, a negative enough image, and a traumatic enough physical threat to disrupt every aspect of a person's life. It also has the capacity to make us uncomfortable enough that we can confront our priorities. Once a person is diagnosed, cancer becomes part of your life; it is not your life. Yes, there are times were it is right "in your face" and unavoidable such as when there are doctor appointments or treatment decisions. At other times, it should recede while you begin reliving your life. Reconnecting with life affirming activities and people is the best way to keep cancer in its place. If work gives you meaning, then it may be important for you. If work was too important and you neglected your family, then more focus might go there. If being more artistic or creative was something you always wanted to do, maybe you could take up painting or a musical instrument.

I often tell people who are diagnosed with cancer that **"You do not have to become a professional cancer patient."** There is more to life than dealing with cancer and a person is certainly much more than their disease. Each person has many facets to their being. Some of the most important areas are listed below. On a scale of 1-10 (1 = little impact, 10 = major impact), identify how great an impact cancer had at the following times since your diagnosis.

Factor	Diagnosis	End Treatment	Six Months Later	One Year Onward
Health/Physical				
Family				
Vocational				
Spiritual				
Emotional				
Sensual/Sexual				
Social				
Recreational				
Living Situation				
Practical Matters				

How has cancer changed your life? Has it all been bad? Some people report that cancer brought them "gifts." They grew closer to family. They learned how much others cared. They made positive changes to reduce stress and make changes they would have never considered without the trauma of cancer in their lives. Researchers are supporting this notion. Several recent studies have looked at how cancer patients have made healthy lifestyle changes such as quitting smoking, changing their diet, and exercising. Researchers use terms like "post-traumatic growth," "benefit finding" and "cancer wellness." This is not to discount the terrible ordeal that cancer can be, but it also may force people to re-evaluate their priorities and redirect their lives.

If work gives you meaning, then it may be important for you. If work was too important and you neglected your family, then more focus might go there. If being more artistic or creative was something you always wanted to do, maybe you could take up painting or a musical instrument.

Health/Physical	1) 2) 3) 4)
Family	1) 2) 3) 4)
Vocational	1) 2) 3) 4)
Spiritual	1) 2) 3) 4)
Emotional	1) 2) 3) 4)
Sensual/Sexual	1) 2) 3) 4)
Social	1) 2) 3) 4)
Recreational	1) 2) 3) 4)
Living Situation	1) 2) 3) 4)
Practical	1) 2) 3) 4)

HOPE: THE LAST WORD

"In the face of uncertainty, there is nothing wrong with hope."
—O. Carl Simonton

Many people suggest that it is wrong to encourage "false hope." **As poet Kay Ryan asks in the poem, "Hope," "What's the use/ of something / as unstable / and diffuse as hope— "** But by its very nature, there is no such thing. The future is always uncertain and is open to possibilities. Hope is a leap of faith and as such is not necessarily rational. One of the few things we know for certain in psychology is that when people lose hope, they give up. That is, they stop trying. More than anything, cancer patients need encouragement to continue living as high a quality of life for as long as possible. There comes a time when our "heart" for living gives out and we face death. Until that point, it is important to live as fully as possible. Full lives require hope. **Without hope of some sort, we literally stop trying. We always need something to look forward to, even if it is for the simple pleasure of savoring a cool drink of water, the gentle touch of a loved one applying balm to parched lips, or the warmth of the sunlight dancing on our skin.**

Even in the direst of circumstances, hope can be maintained. In the concentration camps, Victor Frankl observed that the prisoners who maintained hope while searching for meaning and connection seemed to be the ones who frequently survived. He noted that the last of the human freedoms was "to choose one's attitude in any given circumstances, to choose one's own way." Hope enables us to do that.

Having hope is as necessary as breath and keeps us inspired and open to the future. Expecting a guaranteed or specific outcome is where hope can become false. We may become fixated on a physical cure and close off other avenues of healing, such as resolving interpersonal conflicts with family. We can become so fearful of death or cancer recurrence that we fail to pass on our values. It is unfortunate to lose opportunities for healing.

Hope may come in many forms and it may change over time. When first diagnosed, we certainly hope for a physical cure. Our heart and head may be ready for the fight of our lives. Somewhere along the way, our heart may no longer believe that and our hope may turn to finishing a certain task or being made comfortable before we die. Hope may consist of coming to some understanding of what cancer means in our life or those around us.

As Vaclav Havel states, "Hope is an orientation of the spirit, an orientation of the heart. It is not only the conviction that something will turn out well, but the certainty that something makes sense regardless of how it turns out."

I'm not sure that meaning makes it all okay, but meaning certainly reduces suffering. People try to understand why bad things happen to them. If a trauma has some meaning, then we didn't go through it for nothing. It has some value. David Mamet humorously notes that in his play, *The Duck Variations*:

George: Also they've found a use for cancer.
Emil: Knock wood
George: It's about time. All the millions we spend on research, cigarettes...

Sometimes people also wonder why good things happen to them such as when one of my breast cancer patients expressed her "survivor guilt" upon hearing of the death of another breast cancer patient who had done "everything right" in her battle.

Recent studies have begun to explore the positive side of cancer. They have used terms like "benefit finding," and "post-traumatic growth." I have heard patients use the terms "gifts" or "opportunities" or "wake-up call." They often find out they were really loved a lot more than they had realized or stronger than they ever suspected. They wouldn't have volunteered to contract cancer, but they learned something from it and got not only healthier in body but "weller" than they had been. Some people never feel that anything positive came from cancer and that is very legitimate, too.

We have tried to study hope and hopelessness scientifically. An early study in 1971 by Schmale and Iker reported a relatively high degree of accurate prediction on 68 women with abnormal pap smears who were awaiting biopsy. The researchers interviewed the women and identified hopeful or hopeless women. They predicted 68% of women who had cancer based on hopelessness. Based on hopefulness they predicted 77% of the women who were cancer free. Hopelessness was also related to the promotion of cervical cancer in a 1986 study by Goodkin, Antoni, and Blaney which appeared in the *Journal of Psychosomatic Research*. Some of the same researchers found increased optimism, self-confidence, and emotional expression resulted in less severe dysplasia (which is more likely to lead to cervical cancer).

In a 1996 article in *Psychosomatic Medicine* (58: 113-121) entitled "Hopelessness and Risk of Mortality and Incidence of Myocardial Infarction and Cancer," S. A. Everson, D. E. Goldberg, G. A. Kaplan, and colleagues discuss the relationship of hopelessness in these two major health problems. They suggest hopelessness may play a role in other diseases or health problems as well. We are even looking at brain activity to discover how hope can be manifest. L. A. Gottschalk, J. Fronczek, and M. S. Buchsbaum (*Psychiatry*, 1993, Vol. 56, 270-281) wrote an article called "The Cerebral Neurobiology of Hope and Hopelessness" reflecting the outcomes of a study in which PET scans (Positron Emission Tomography) were used to examine the functional activity of regional brain glucose metabolism in 10 normal men. This analyzed their thinking and word content for hope and hopelessness while they imagined designated events in their own lives. Specific areas of the brain were activated in those with hopeful versus more hopeless scores. In earlier work Gottschalk found that hopefulness in 36 patients getting radiation for advanced cancer was related to increased survival.

While these studies are fascinating, we may ultimately never fully understand the mechanism behind hope and it may remain somewhat of a mystery. As Emily Dickinson relates:

Hope is a Thing with feathers
that perches in the Soul
and Sings the tune
without the Words
and never stops at all.

False hope is generated only when we become too fixated on our own agenda and lose touch with our larger connection or purpose for being here. I am reminded of Bernie Siegel's bumper sticker that reads, "Everything in the Universe is subject to change and everything is on schedule." Hopefully, this book has given you some tools to help you focus on what you can control. May it make your path toward wellness easier and keep you on schedule.

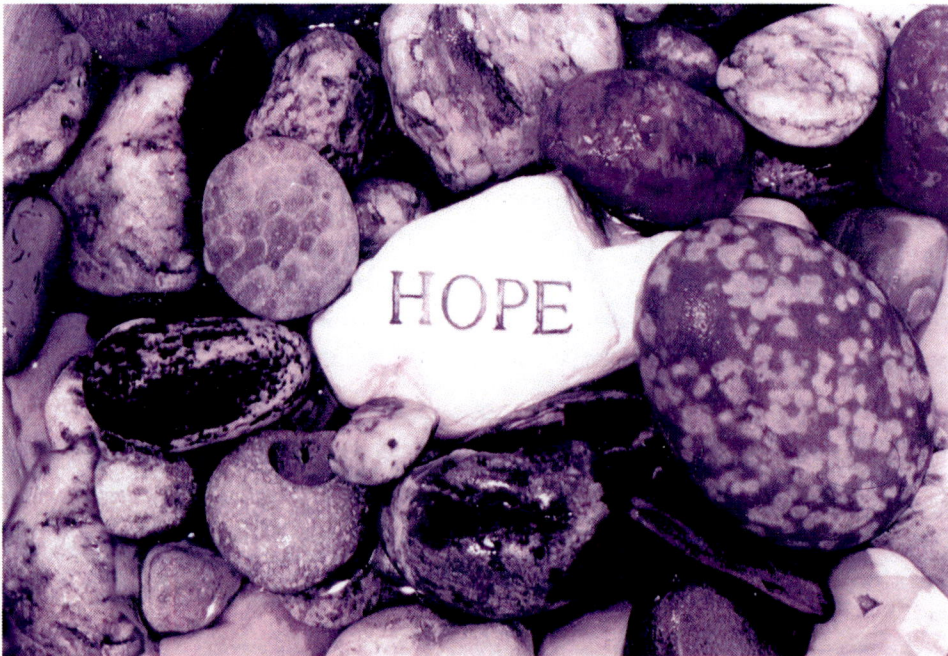

Photo: Courtesy of Barbara Armstrong

References
and
Resources

A FEW GOOD REFERENCES
FOR STRESS MANAGEMENT

Alberti, R.E., & Emmons, M.L., *Stand Up, Speak Out, Talk Back!,* New York: Pocket Books, 1975.

Benson, H., *The Relaxation Response,* New York: Avon, 1975.

Bernhardt, R., & Martin, D., *Self-Mastery through Self-Hypnosis,* New York: Signet, 1977.

Bloomfield, H.M., & Cooper, R.K., *The Power of 5,* Emmaus,PA: Rodale Press, 1995.

Burns, D.D., *Feeling Good: The New Mood Therapy,* New York: Avon, 1999.

Charlesworth, E.A., & Nathan, R.G., *Stress Management,* New York: Ballantine Books, 1984.

Davis, M., Eshelman, E.R., & McKay, M., *The Relaxation and Stress Reduction Workbook,* New York: MJF Books, 1995.

Ellis, A., *A New Guide to Rational Living,* North Hollywood, CA: Wilshire Books, 1975.

Fensterheim, H. & Baer, J., *Don't Say Yes When You Want to Say No,* New York: Dell Publishing Co., 1975.

Hadley, J. & Staudacher, C., *Hypnosis for Change,* New York: Ballantine, 1985.

Hunt, D., & Mait, P., *The Tao of Time,* New York: Simon & Schuster, 1990

Kabat- Zinn, J., *Full Catastrophe Living: Using the Wisdom of Your Body and Mind to Face Stress, Pain, and Illness,* New York: Delta Books, 1991.

Kabat- Zinn, J., *Wherever You Go, There You Are Mindfulness Meditation in Everyday Life,* New York: Hyperion, 1994.

Lakein, A., *How to Get Control of Your Time and Your Life,* New York: Signet Books, 1973.

Mason, L. J., *Guide to Stress Reduction,* Berkeley, CA: Celestial Arts, 1985.

Mason, L.J., *Stress Passages,* Berkeley, CA: Celestial Arts, 1988.

Miller, L.H., *The Stress Solution,* New York: Pocket Books, 1993.

Nathan, R.G., Staats, T.E., & Rosch, P.J., *The Doctors' Guide to Instant Stress Relief,* New York: Ballantine Books, 1987.

Pelletier, K., *Mind as Healer, Mind as Slayer*, New York: Delta Books, 1977.

Pennebaker, J.W., *Opening Up*, New York: William Morrow and Company, Inc., 1990.

Scott, D., *How to Put More Time in Your Life*, New York: Signet, 1980.

Shaffer, M., *Life after Stress*, Chicago, IL, Contemporary Books, Inc., 1983.

Stroebel, C.F., *QR The Quieting Response*, New York: Berkley Books, 1982.

Woolfolk, R.L. & Richardson, F.L., *Stress, Sanity, and Survival*, New York: Signet, 1978.

Add new stress management books that you find helpful here:

_____

_____

_____

_____

_____

_____

_____

_____

_____

_____

_____

_____

_____

_____

_____

_____

A FEW SELECTED CANCER REFERENCES

Achterberg, J., Dossey, B., & Kolkmeier, L., *Rituals of Healing*, New York: Bantam Books, 1994.

Achterberg, J., & Lawlis, G.F., *Imagery and Disease*, Champaign, IL: Institute for Personality and Ability Testing, Inc., 1984.

Anderson, G., *50 Essential Things to Do When the Doctor Says It's Cancer*, New York: Plume, 1993.

Armstrong, L. & Jenkins, S., *It's Not about the Bike: My Journey Back To Life*, Berkley Pub Group, 2001

Babcock, E.N., *When Life Becomes Precious*, New York: Bantam Books, 1997.

Benjamin, H., *Wellness Community Guide to Fighting for Recovery from Cancer*, New York: Putnam, 1995.

Benson, H., *Timeless Healing*, New York: Fireside-Simon & Schuster, 1996.

Berman, J., Fleegler, F., & Hanc, J., *The FORCE PROGRAM: The Proven Way to Fight Cancer Through Physical Activity and Exercise*, New York: Balentine Books, 2001.

Bock, K., and Sabin, N., *The Road to Immunity*, New York: Pocket Books, 1997.

Bognar, D., *Cancer Increasing Your Odds for Survival*, Alameda, CA: Hunter House, 1998.

Bolen, J. S., *Close to the Bone*, New York: Scribner, 1996.

Buchholz, B. & Buchholz S., *Live Longer, Live Larger: A Holistic Approach for Cancer Patients and Families*, Sebastopol, CA, 2001.

Canfield, J., Hansen M.V., Aubery, P., & Mitchell, N., *Chicken Soup for the Surviving Soul*, Deerfield Beach, FL, 1996.

Cicala, R.S., *The Cancer Pain Sourcebook*, Lincolnwood, IL: Contemporary Books, 2001.

Cotter, A., *from this moment ON*, New York: Random House, 1999.

Cousins, N., *Head First: The Biology of Hope*, New York: E.P. Dutton, 1989.

Cukier, D., & McCullough, V.E., *Coping with Radiation Therapy,*
Los Angeles, CA: Lowell House, 1993.

Diamond, W. J., Cowden, W. L., & Goldberg, B., *An Alternative Medicine Definitive Guide to Cancer*, Tiburon, CA: Future Medicine Publishing, Inc., 1998.

Dollinger, M., Rosenbaum, E.H., Tempero, M., & Mulvihill, S.J. *Everyone's Guide to Cancer Treatment* (Fourth Edition), Toronto, Ontario, Canada: Somerville House Books Limited, 2002.

Dossey, L., *Prayer Is Good Medicine,* New York: HarperCollins, 1996.

Dreher, H., *The Immune Power Personality*, New York: Plume, 1995.

Gaynor, M.L., *Healing Essence: A Cancer Doctor's Practical Program for Hope and Recovery,* New York: Kodansha, 1995.

Gaynor, M.L., *Sounds of Healing*, New York: Broadway Books, 1999.

Gaynor, M.L., and Hickey, J., *Dr. Gaynor's Cancer Prevention Program, New York:* Kensington Books, 1999.

Geffen, J., *The Journey through Cancer*, New York: Crown Publishers, 2000.

Gersh, W.D., Golden, W.L., & Robbins, D.M., *Mind over Malignancy,* Oakland, CA: New Harbinger, 1997.

Girard, V., *There's No Place Like Hope*, Lynnwood, CA: Compendium, Inc. 2001.

Hess, D.J. (ed.) *Evaluating Alternative Cancer Therapies*, Piscataway, NJ: Rutgers University Press, 1999.

Hirshberg, C., & Barasch, M.I., *Remarkable Recovery,* New York: Riverhead Books, 1995.

Holland, J., & Lewis, S., *The Human Side of Cancer: Living with Hope, Coping with Uncertainty*, Quill, 2001.

Jordan, H., *No Such Thing as a Bad Day* , Marietta, GA : Longstreet Press, 2000.

King, D., King, J., & Pearlroth, J., *Cancer Combat,* New York: Bantam Books, 1998.

Kramp, E.T., Kramp, D.H., & McKhann, E.P., *Living with the End in Mind,* New York: Three Rivers Press, 1998.

LeShan, L., *You Can Fight for Your Life*, New York: Evans, 1979.

LeShan, L., *Cancer as a Turning Point: A Handbook for People with Cancer, Their Families, and Health Professionals*, New York: E.P. Dutton, 1989.

Lerner, M., *Choices in Healing: Integrating the Best of Conventional and Complementary Approaches to Cancer*, Cambridge, MA: The MIT Press, 1994.

LeVert, S., *When Someone You Love Has Cancer*, New York: Dell Publishing, 1995.

Morra, M., & Potts, E., *Choices: Realistic Alternatives in Cancer Treatment*, New York: Avon, 1996.

Ornish, D., *Love and Survival*, New York:HarperCollins, 1998.

Pearsall, P., *Making Miracles*, New York: Prentice Hall Press, 1991.

Prasad, K.D., *Vitamins in Cancer Prevention and Treatment A Practical Guide*, Rochester, VT: Healing Arts Press, 1994.

Quillin, P., & Quillin, N., *Beating Cancer with Nutrition*, Carlsbad, CA Nutrition Times Press, Inc., 2001.

Quillin, P., & Quillin, N., *Beating Cancer with Nutrition*, Tulsa, OK: Nutrition Times Press, Inc., 1994.

Siegel, B., *Love, Medicine and Miracles*, New York: Harper & Row, 1986.

Siegel, B., *Peace, Love, and Healing*, New York: Harper & Row, 1989.

Siegel, B., *How to Live between Office Visits*, New York: Harper & Row, 1993.

Simon, D., *Return to Wholeness*, New York: John Wiley & Sons, Inc., 1999.

Simone, C.B., *Cancer & Nutrition*, Garden City Park, NY: Avery Publishing Group, Inc., 1992.

Simonton, O.C., Matthews-Simonton, & Creighton, J.L., *Getting Well Again*, New York: Bantam Books, 1992.

Smith, G.W., & Naifeh, S., *Making Miracles Happen,* Boston, MA,: Little, Brown and Company, 1997.

Spiegel, D., *Living Beyond Limits: New Hope for Facing Life-Threatening Illness,* New York: Times Books, 1993.

Temoshok, L., & Dreher, H., *The Type C Connection: The Behavioral Links to Cancer and Your Health,* New York: Random House, 1992.

Willis, J., *The Cancer Patient's Workbook : Everything You Need to Stay Organized and Informed,* New York: Dorling Kindersley Publishing Co., 2001

Yance, D.R. Jr., & Valentine, A., *Herbal Medicine, Healing & Cancer,* Lincolnwood (Chicago), IL: Keats Publishing, 1999.

Add any new useful cancer resource books you find here:

_____

_____

_____

_____

_____

_____

_____

_____

A FEW HELPFUL INTERNET SITES

Because the Internet changes so quickly, I am listing only a limited number of key sites. These are just a few of the sites my patients have found useful. I have not listed chat rooms or online support groups, as those are a matter of individual taste. I have also just listed general sites as opposed to specific cancer sites. That would have made the list vastly longer. Another caution is the entrepreneurial nature of the Internet: many people are out to sell commercial products and may take advantage of people who are desperate. Be careful to verify whether a site is reputable and secure.

AMERICAN CANCER SOCIETY —http://www.cancer.org
AMERICAN INSTITUTE FOR CANCER RESEARCH
                  —http://www.airc.org
AMERICAN SOCIETY OF CLINICAL ONCOLOGY(ASCO)
                  —http://www.asco.org
R.A. BLOCH CANCER FOUNDATION —http://www.blochcancer.org,
                  —http://www.blastcancer.org
CANCER CARE, INC. —http://www.cancercare.org
CANCER EDUCATION —http://www.canceducation.com
CANCER HOPE NETWORK —http://www.cancerhopenetwork.org
CANCER LINKS —http://www.cancerlinks.com

CANCERNET —http://wwwicic.nci.nih.gov/
CANCER NEWS ON THE NET —http://www.cancernews.com/quickload.htm
CANCERLIT —http://wwwicic.nci.nhi.gov/canlit/canlit.htm
CANCER RESEARCH INSTITUTE —http://www.cancerresearch.org
CANSEARCH Websites —www.cansearch.org/canserch/canserch.htm
CANSEARCH —http:www.access.digex.net/~mkragen/cansearch.html
CENTER FOR ADVANCEMENT IN CANCER EDUCATION
                  —www.lifeenrichment.com/cancer.htm
CHEMO CARE —www.chemocare.com
CLINICAL TRIALS — http://clinicaltrials.gov/ct/gui/c/b
COALITION OF NATIONAL CANCER COOPERATIVE GROUPS —www.cancertrialshelp.org
COPING WITH CANCER MAGAZINE —www.copingmag.com
CORPORATE ANGEL NETWORK, INC.
                  —http://www.corpangelnetwork.org
GILDA'S CLUB WORLDWIDE —www.gildasclub.org
INTERNATIONAL CANCER ALLIANCE —http://www.icare.org/icare
MEDSCAPE ONCOLOGY HOME PAGE
—http://www.medscape.com/Home/Topics/oncology/oncology.html
NATIONAL CANCER INSTITUTE —http://www.nci.nih.gov
NATIONAL CENTER FOR COMPLEMENTARY AND ALTERNATIVE MEDICINE
          —http://nccam.nih.gov/nccam
NATIONAL COALITION FOR CANCER RESEARCH
                  —http://www.cancercoalition.com

NATIONAL COALITION FOR CANCER SURVIVORSHIP
    —http://www.canceradvocacy.org
NATIONAL COMPREHENSIVE CANCER CENTER NETWORK —http://www.nccn.org
NATIONAL HOSPICE ORGANIZATION —http://www.nho.org/general.htm
ONCOLINK : U of PA —www.oncolink.upenn.edu
PATIENT ADVOCATE FOUNDATION
                            —http://www.patientadvocate.org
ALL CANCER LISTS -- http://cure.medinfo.org
STEVE DUNN'S CANCER INFORMATION PAGE
                            —http://www.cancerguide.org/
PNIRS -- http://www.urmc.rochester.edu/PNIRS/welcome.html
Ralph W. Moss, Ph.D. —cancerdecisions.com
STRESS AND IMMUNITY BREAST CANCER PROJECT
    —http://www-cancer.med.ohio-state.edu/clinical/ADMIN/brstrss.htm
PNI GRAPHICS —http://web.indstate.edu/nurs/bennett/pni2.htm
THE WELLNESS COMMUNITY —http://www.wellness-community.org

SOME IMPORTANT CANCER RESOURCES

Below are listed the important addresses and telephone numbers of valuable cancer resources. Space does not permit listing all the resources available. You can add important resources that you specifically need at the end of this section or on the back of these pages. Web pages for some of these resources are listed in the internet section.

American Cancer Society (ACS), 1599 Clifton Road, NE, Atlanta, GA 30329, 800-ACS-2345

American Institute for Cancer Research (AICR), 1759 R Street, NW, Washington, D.C. 20009, 800-843-8114 or 202-328-7744 (In D.C. area)

R.A. Bloch Cancer Foundation, Inc., 4435 Main Street, Kansas City, MO 64111, 800-433-0464, 816-932-8453

Cancer Care, Inc., 1180 Avenue of the Americas, New York, NY 10036, 800-813-HOPE, 212-302-2400

Cancer Hope Network, 2 North Road, Suite A, Chester, NJ 07930, 877-HOPENET (toll free), 908-879-4039

Cancer Research Institute, 681 Fifth Avenue, New York, NY 10022, 800-99-CANCER, 212-688-7515

Coping with Cancer (magazine), P.O. Box 682268, Franklin, TN 37068-2268, 615-790-2400

International Cancer Alliance, 4853 Cordell Avenue, Suite 11, Bethesda, MD 20814, 800-ICARE-61

Make Today Count, c/o Connie Zimmerman, St. John's Mid- America Cancer Center, 1235 E. Cherokee, Springfield, MO 65804-2263, 800-432-2273

National Cancer Institute (NCI), Cancer Information Service 800-4-CANCER

National Coalition for Cancer Research (NCCR), 426 C Street NE, Washington, DC 20002, 202-544-1880

National Coalition for Cancer Survivorship (NCCS), 1010 Wayne Avenue, Silver Spring, MD 20910, 888-937-6227

National Hospice Organization (NHO),1901 N. Moore Street, Suite 901, Arlington, VA 22209, 800-658-8898, 703-243-5900

Oncolink, University of Pennsylvania Cancer Center, 3400 Spruce Street, 2 Donner, Philadelphia, PA 19104, 215-349-8895

Patient Advocate Foundation (PAF), 780 Pilot House Drive, Suite 100C, Newport News, VA 23606, 800-532-5274, 757-873-6668

The Wellness Community, 35 East 7th Street, Suite 412, Cincinnati, OH 45202 888-793-WELL, 513-421-7111

DAY-BY-DAY

Monthly Cancer Wellness Journal ©

S	M	T	W	T	F	S

HOW TO USE THIS MEDITATIVE JOURNAL

Life has gotten incredibly complex and fast. It is easy to get overwhelmed and fall into patterns of unconscious reactivity instead of making conscious choices. Cancer makes it even easier to loose our grounding and our center. Making a conscious choice to slow down and listen to our inner voice is healthy. Slowing down heals our souls and helps us understand what is happening to us. It is helpful to combine our intuitive (inner) wisdom with the rational knowledge we gather to make our best decisions. This combination may have a beneficial effect on our physical body as well. This meditative journal is a way to quiet the endless static and really listen to the wisdom we all have from our connection to the Source of the Universe.

Whether you choose to use the structure supplied here or develop your own, the daily practice of becoming quiet in order to slow things down will be beneficial. You will need to find a quiet place that can be conducive to peacefulness. It can be the same place every time or you may vary it if you like. Try to schedule this at the same time everyday, so the routine will be easier. People who vary the time too much, often find it gets "lost in the shuffle." Start with just a small amount of time and build up as you can. Develop more conscious awareness throughout the day. If you prefer something more active, a walking meditation might be good. Many people like to begin the day by reading their passage. It can then act as the "Thought of the Day" and help you look at things consciously. Initially, it might not make much impact, but with practice, you will become more aware. You may want to begin saving your own inspiring quotations and use those as your meditation. There is a blank format included for you to develop your own journal. Reading and thinking about it may be enough. There is also a blank calendar to mark important dates and appointments. You may copy both forms as much as you would like.

Journaling is a way to process what is happening to us and help us understand what we are going through. Not everyone needs or likes to write. But many feel that writing can make experiences more real and help sort through them more easily. Whether you choose to write or not, hopefully you will enjoy some of the quotations and affirmations and adopt them as your own. A two-month sample follows. We are in the process of developing a yearlong book. Hopefully, you will consider it a lifelong process.

"Fall seven times, stand up eight."
—Japanese Proverb

A chronic illness such as cancer is a roller coaster. There are certainly ups and downs. There is good news and there are setbacks. There are many sources of strength upon which to draw. These "sources" help to lift us back up as long as we are willing to access them. We may also have more strength than we credit ourselves. Today, I will focus on what motivates me to get back up should I fall down.

Your thoughts, feelings, and observations today:

_____

_____

_____

_____

_____

_____

_____

_____

Today's Affirmation

Today, I will acknowledge my deepest sources of strength and motivation.

My Personal Affirmation

_____

_____

"There are two ways to live your life. One is as though nothing were a miracle. The other is as though everything is a miracle."
—Albert Einstein

Discussion

_____
_____
_____
_____
_____

Your thoughts, feelings, and observations today:

_____
_____
_____
_____
_____
_____
_____
_____

My Personal Affirmation

_____
_____

"You are not a professional cancer patient. That is not your occupation"
—Morry Edwards, Ph. D.

There are many facets to being a human being. We are more than just a profession or occupation. We are also more than a spouse, a child, a religion, or a club member. There is much of us to explore. There are times when cancer comes to the forefront of our attention and we need to deal with it. But there are many more times when we need to let cancer recede into the background. Today, I will let cancer fade into the background and explore some of my other selves.

Your thoughts, feelings, and observations today:

_____

_____

_____

_____

_____

_____

_____

_____

_____

Today's Affirmation

I will appreciate the many aspects of my being.

My Personal Affirmation

_____

_____

"He (she) who laughs, lasts."
—Mary Pettibone Poole
*"Life does not cease to be funny when people die
any more than it ceases to be serious when people laugh."*
—George Bernard Shaw

✠

Discussion

_____

_____

_____

_____

Your thoughts, feelings, and observations today:

_____

_____

_____

_____

_____

_____

_____

My Personal Affirmation
✠

_____

_____

"The cancer patient who keeps up a false front in the name of 'positive attitude' is doing him(her)self a disservice. He's (She's) cutting himself (herself) off from emotions — fear, anger, sadness — that are necessary in the healing process."
—Lydia Temoshok

It takes a great deal of energy to project a mask to the outside world. Even without having cancer this is wasteful. It is like heat going up your chimney instead of warming your house. Once cancer is diagnosed, it is important to mobilize our energy and direct it toward healing. To encourage the healing process, we need to allow ourselves to have a full range of feelings that occur in that situation. Having feelings like anger or sadness are normal and will not hurt me or take away from my "positive attitude," unless I suppress them. Today, I will explore my real feelings and project an honest face.

Your thoughts, feelings, and observations today:

_____

_____

_____

_____

_____

_____

Today's Affirmation

I will project an honest face, so I will not waste my energy.

My Personal Affirmation

_____

_____

"Since everything is in our heads, we had better not lose them."
—Coco Chanel

✠

Discussion

_____

_____

_____

_____

Your thoughts, feelings, and observations today:

_____

_____

_____

_____

_____

_____

_____

My Personal Affirmation

✠

_____

_____

"My present purpose is... to state a fact—that cancer, even when advanced in degree and of long duration, may get better, and does sometimes get well. There is a cure of cancer, apart from operative removal... These cases... are the sun of our hope."
—Sir Alfred Pearce Gould (1910)

At least one person has survived every form of cancer. Even though a certain diagnosis may not carry a large percentage of cure, there is always reason to hope for a positive outcome. No one can inspire more hope than another person who has been through the fire of cancer and come through stronger. Consider attending a support group or seek out a cancer survivor for encouragement.

Your thoughts, feelings, and observations today:

_____

_____

_____

_____

_____

_____

Today's Affirmation

I am not alone. There is someone else who has survived my type of cancer and lived a full life.

My Personal Affirmation

_____

_____

"Pay attention and you'll find many perfect moments in any given day."
—Steven C. Paul, *Inneractions*

�֍

Discussion

_____

_____

_____

_____

_____

Your thoughts, feelings, and observations today:

_____

_____

_____

_____

_____

_____

_____

_____

My Personal Affirmation

✖

_____

_____

"There is the type of person in whose mind
God is always mixed up with vitamins."
—Manly P. Hall

✦

There are many factors that account for health or illness. As yet we cannot establish a precise equation to account for the exact percentage of everything that helps us into wellness. Our responsibility is to seek out that which benefits us. Today, I will explore new avenues that potentially could help me.

Your thoughts, feelings, and observations today:

_____

_____

_____

_____

_____

_____

_____

_____

_____

Today's Affirmation

I will explore all avenues of healing
that feel beneficial to me.

My Personal Affirmation

✦

_____

_____

"If God is your partner, make your plans large."
—Martha Lupton

✠

Discussion

_____

_____

_____

_____

_____

Your thoughts, feelings, and observations today:

_____

_____

_____

_____

_____

_____

_____

_____

My Personal Affirmation
✠

_____

_____

"The purpose of life is a life of purpose."
—Robert Byrne

It is natural and important to question why we are here. Often times when a person is diagnosed with cancer, that search becomes even more poignant. We need goals or purposes for our lives. It serves to focus our energy and create a sense of enthusiasm and accomplishment. Don't be afraid that getting cancer means you no longer set goals. Instead goals can be a way to increasing the quality and possibly the quantity of your life. Today, I will spend time focusing on what is my purpose in life.

Your thoughts, feelings, and observations today:

_____

_____

_____

_____

_____

_____

_____

_____

Today's Affirmation

Developing a purpose in life can increase my strength and healing.

My Personal Affirmation

_____

_____

"Don't deny the diagnosis. Try to defy the verdict."
—Norman Cousins

✦

Discussion

_____

_____

_____

_____

_____

Your thoughts, feelings, and observations today:

_____

_____

_____

_____

_____

_____

_____

_____

My Personal Affirmation

✦

_____

_____

"Sometimes I sits and thinks and sometimes I just sits."
—Satchel Paige

�֍

The world moves mighty fast at times. We all need down time or as a friend of mine put it, "Veg Time". Sometimes we need to meditate on a particular question that is facing us and sometimes we need to process what has been presented to us. But sometimes we just need time to recharge and regenerate our energy. And that might mean just staring into space. Today I will make some time to refresh and renew myself.

Your thoughts, feelings, and observations today:

_____

_____

_____

_____

_____

_____

_____

Today's Affirmation

I will allow myself plenty of time to just "sit" and recharge my energy today.

My Personal Affirmation

✖

_____

_____

*"...a little private regime. You just lie there very still, and you just say,
'I've got cancer— get the hell out of here.'"*
—Lesley Bermingham, a cancer patient
Described in *Remarkable Recovery*
by Caryle Hirshberg & Marc Ian Barasch

❖

Discussion

_____

_____

_____

_____

Your thoughts, feelings, and observations today:

_____

_____

_____

_____

_____

_____

My Personal Affirmation
❖

_____

_____

"Woes have their ebb and flow."
—Katherine Fowler Phillips

Our journey through life has its ups and downs. That is especially true after the diagnosis of cancer. One minute we might get good news from a doctor the next we might feel a new suspicious lump. Nothing in particular may cause us to feel down during a certain period. Even though it might not seem like it at the time, these down periods are normal and do pass. It is easy to get support and help when times are good. Today, I will generate some ideas to help support me through the tough periods.

Your thoughts, feelings, and observations today:

_____

_____

_____

_____

_____

_____

_____

_____

Today's Affirmation

My woes will not last forever.

My Personal Affirmation

_____

_____

*"A wide range of feelings and reactions to cancer is normal, but—
I need to redirect my energy from unproductive emotions such as worry, anger,
fear, and resentment into acceptance, love, and healing."*
—Morry Edwards, Ph.D.

❈

Discussion

_____

_____

_____

_____

Your thoughts, feelings, and observations today:

_____

_____

_____

_____

_____

_____

_____

My Personal Affirmation

❈

_____

_____

"Cancer is probably the most unfunny thing in the world, but I'm a comedienne, and even cancer couldn't stop me from seeing humor in what I went through."
—Gilda Radner

�֍

No matter how terrible a situation may be, humor can be an effective way for us to ease our tension and be less frightened. Even though it may be difficult, I will take some time today and learn to laugh again.

Your thoughts, feelings, and observations today:

_____

_____

_____

_____

_____

_____

_____

_____

_____

Today's Affirmation

Laughter can be an important way for me to cope.

My Personal Affirmation

�֍

_____

_____

*"Life is a journey of discovery, not certainty,
and the best way to make it is to simply take it."*
—Dr. Carl Hammerschlag, *The Theft of the Spirit*

✠

Discussion

_____

_____

_____

_____

_____

Your thoughts, feelings, and observations today:

_____

_____

_____

_____

_____

_____

_____

My Personal Affirmation

✠

_____

_____

"Life is what happens to you while you're busy making other plans."
—John Lennon

✵

While it is important to set goals and make plans, we need to be flexible enough to deal with what comes our way. While we look toward the future, we need to detach enough that we can enjoy the now. Today, I will balance making plans for the future and enjoying the moment.

Your thoughts, feelings, and observations today:

_____

_____

_____

_____

_____

_____

_____

_____

_____

Today's Affirmation

*I will not get so attached to my plans
that I don't live in the moment.*

My Personal Affirmation

✵

_____

_____

*"I was thrown out of college for cheating on the metaphysics exam;
I looked into the soul of the boy next to me."*
—Woody Allen

�֍

Discussion

_____

_____

_____

_____

Your thoughts, feelings, and observations today:

_____

_____

_____

_____

_____

_____

My Personal Affirmation

✖

_____

_____

"Some people are so afraid to die that they never begin to live."
—Henry Van Dyke

Being diagnosed with cancer is a scary situation. But if we allow ourselves to get stuck in fear then it can easily destroy our enjoyment of life. We stop making plans and begin waiting for all these horrible thoughts to become reality. Today, I will spend some time staring those fears in the face and trying to focus on constructive actions to reduce those concerns. I will also try and enjoy the present and make positive plans for enjoyable goals in the future as a first step forward in making them happen.

Your thoughts, feelings, and observations today:

Today's Affirmation

I will face my fears today so they do not interfere with my enjoyment of life.

My Personal Affirmation

"Millions long for immortality who do not know what to do with themselves on a rainy Sunday afternoon."
—Susan Ertz

Discussion

_____
_____
_____
_____
_____

Your thoughts, feelings, and observations today:

_____
_____
_____
_____
_____
_____
_____
_____

My Personal Affirmation

_____
_____

"It is good to have an end to journey toward:
but it is the journey that matters in the end."
—Ursula K. LeQuin

※

Goals throughout life are important, but often we get so focused on the end product that we neglect the enjoyment that comes with the process of the doing. In the '60s, it was said that the journey was part of the trip. With cancer, many people stop making goals because they are unsure of the ability to complete them. Today, I will focus on enjoying the strides toward my goal rather than whether I complete it.

Your thoughts, feelings, and observations today:

_____

_____

_____

_____

_____

_____

_____

Today's Affirmation

Today as everyday, I can enjoy the process toward
the goal instead of the end product.

My Personal Affirmation

※

_____

_____

*"You are responsible for your own life
and have a job to perform in your health care."*
—Neil A. Fiore

�֍

Discussion

_____

_____

_____

_____

Your thoughts, feelings, and observations today:

_____

_____

_____

_____

_____

_____

_____

My Personal Affirmation

✖

_____

_____

"Of all sad words of tongue or pen, The saddest are these:
It might have been."
—John Greenleaf Whittier

There is probably no one alive without regrets about his or her life. Many times a person has no impetus to examine life. While not asked for, the diagnosis of cancer provides an opportunity to focus on what is really important. Take this time today to examine any regrets you may have and decide if there is anything you can do about them. Also, work on letting go of any regrets that cannot be changed.

Your thoughts, feelings, and observations today:

_____

_____

_____

_____

_____

_____

_____

_____

Today's Affirmation

I will do all that I can to eliminate regrets
that I have stored through my life.

My Personal Affirmation

_____

_____

173

""May those
whose lives are gripped in the palm of suffering
open
even now
to the Wonder of Life
that is our joyous Unity with Holiness.
—Rabbi Rami M. Shapiro
Prayers for Healing
365 Blessings, Poems, and Meditations from around the World

Your thoughts, feelings, and observations today:

_____

_____

_____

_____

_____

_____

_____

_____

_____

_____

My Personal Affirmation

_____

_____

"There's never enough time to do all the nothing you want."
—Bill Watterson

It is easy to become caught up in the dizzying pace of time and the consuming thought of being productive every minute. Especially, when one is diagnosed with cancer and feels that this time on earth is really limited. Remember only you can truly slow things down and savor your experiences. Today, I will make a point of taking time to do nothing except fully experience what happens to me.

Your thoughts, feelings, and observations today:

_____

_____

_____

_____

_____

_____

_____

_____

Today's Affirmation

Today and every day I will savor my experiences.

My Personal Affirmation

_____

_____

"That which does not kill me, makes me stronger."
"He who has a why to live for can bear almost any how."
—Friedrich Nietzche

�֎

Discussion

_____

_____

_____

_____

Your thoughts, feelings, and observations today:

_____

_____

_____

_____

_____

_____

_____

My Personal Affirmation
�֎

_____

_____

"I no longer prepare food or drink with more than one ingredient."
—Cyra McFadden

There is a lot of truth to the sentiment, "You are what you eat." While you may not feel this sentiment to such an extreme, eating more nutritionally may be a beneficial tool in your wellness program. Today, I will spend some time examining my diet and foods that make me feel healthier.

Your thoughts, feelings, and observations today:

_____

_____

_____

_____

_____

_____

_____

_____

_____

Today's Affirmation

I will take steps to healthier eating
as a part of my wellness program.

My Personal Affirmation

_____

_____

"To be upset over what you don't have is to waste what you do have."
—Ken S. Keyes, Jr.

Discussion

_____

_____

_____

_____

_____

Your thoughts, feelings, and observations today:

_____

_____

_____

_____

_____

_____

_____

My Personal Affirmation

_____

_____

"If I had known I was going to live this long
I would have taken better care of myself."
—Unknown (from Robert Byrne's *1,911 Best Things Anybody Ever Said)*

No matter how long or short we live, it is never too late to take the very best care of yourself that you can. It is important that your heart is in the changes that you make and that you make them joyfully. Today, I will focus on how I can take better care of myself and put those changes into action.

Your thoughts, feelings, and observations today:

_____

_____

_____

_____

_____

_____

_____

_____

Today's Affirmation

It is always important to take the best care of myself.

My Personal Affirmation

_____

_____

"Merely to have survived is not an index of excellence."
—*The New Union Prayer Book*
Surviving is important. Thriving is elegant."
—Maya Angelou
Hallmark Greeting

Discussion

_____
_____
_____
_____

Your thoughts, feelings, and observations today:

_____
_____
_____
_____
_____
_____
_____

My Personal Affirmation

_____
_____

"Be patient with everyone, but above all with yourself."
—St. Francis DeSales

We live in a fast paced world and want instant gratification of our wants. Sometimes it is healthier for us to slow down and get in touch with our genuine needs. Often, too, our impatience and struggle with the world is a reflection of our impatience and self-critical attitude with ourselves. Today I will take time to slow down and tune in to what I really need.

Your thoughts, feelings, and observations today:

Today's Affirmation

I will slow down to learn my genuine needs today.

My Personal Affirmation

"The only handicap is a bad attitude."
—Scott Hamilton

Discussion

Your thoughts, feelings, and observations today:

My Personal Affirmation

"True silence is the rest of the mind;
it is to the spirit what sleep is to the body, nourishment and refreshment."
—William Penn

Life is noisy and particularly the sound of all the thoughts rushing through our heads. When we are busy in our daily lives and interacting socially, it is often difficult for us to hear our true feelings and needs. It is beneficial for us to cultivate silence for a brief time on a regular basis, so that we can be in contact with our true thoughts. Today, I will make a point of cultivating silence.

Your thoughts, feelings, and observations today:

_____

_____

_____

_____

_____

_____

_____

_____

Today's Affirmation

I will use silence to connect with my true self.

My Personal Affirmation

_____

_____

183

"I try to take one day at a time,
but sometimes several days attack me at once."
—Ashleigh Brilliant

�֎

Discussion

_____

_____

_____

_____

_____

Your thoughts, feelings, and observations today:

_____

_____

_____

_____

_____

_____

_____

_____

My Personal Affirmation
�֎

_____

_____

"It is very inconvenient to be mortal—
You never know when everything may suddenly stop happening."
—Ashleigh Brillant

Being diagnosed with a chronic or life-threatening disease like cancer really calls attention to our mortality. It also reminds us about the uncertainty of this life. Someone once remarked, "Life is uncertain, so eat dessert first." Today, I will concentrate on what is really important and not procrastinate dealing with those priorities.

Your thoughts, feelings, and observations today:

_____

_____

_____

_____

_____

_____

_____

_____

Today's Affirmation

I accept life's uncertainty
and will act in harmony with my priorities.

My Personal Affirmation

_____

_____

*"Regret is an appalling waste of energy; you can't build on it;
it is only good for wallowing."*
—Katherine Mansfield

Discussion

_____
_____
_____
_____

Your thoughts, feelings, and observations today:

_____
_____
_____
_____
_____
_____
_____
_____

My Personal Affirmation

_____
_____

"The change of one simple behavior
can affect other behaviors and thus change many things."
—Jean Baer

Change is difficult and it may be especially difficult during the turmoil of cancer. We often look to change things and find so many things to change that we get overwhelmed. It might be wise to start with one small, specific area where we can be successful rather than just setting an arbitrary goal. Success is motivating and will help us gain confidence to tackle the next area and the next. Just remember to gear for success and get one area under control at a time.

Your thoughts, feelings, and observations today:

_____

_____

_____

_____

_____

_____

_____

_____

Today's Affirmation

I will regain control of my life
by making one small change.

My Personal Affirmation

_____

_____

*"The way I see it, if you want the rainbow,
you gotta be willing to put up with the rain.*
—Dolly Parton
"You can't have a testimony without a test.
—Iyanla Vanzant

�֎

Discussion

_____

_____

_____

_____

Your thoughts, feelings, and observations today:

_____

_____

_____

_____

_____

_____

_____

_____

My Personal Affirmation
✖

_____

_____

"There are only three important activities in life: to learn and understand what we can, to enjoy ourselves without harming others, and to share love."
——Morry Edwards, Ph.D.

Cancer is a truly intrusive force in a person's life. The occurrence of cancer invades every aspect of a person's being from physical intimacy through social relationships; from job security to changes in food preference. One thing cancer can also do is cause us to focus on what is most important in our lives and help us make changes in that direction. Today provides an opportunity for me to prioritize what is important in my life.

Your thoughts, feelings, and observations today:

_____

_____

_____

_____

_____

_____

_____

_____

Today's Affirmation

I will keep in mind what is truly important.

My Personal Affirmation

_____

_____

"Life is not a matter of having good cards, but of playing a poor hand well."
—Robert Louis Stevenson

�֎

Discussion

_____

_____

_____

_____

_____

Your thoughts, feelings, and observations today:

_____

_____

_____

_____

_____

_____

_____

_____

My Personal Affirmation
✖

_____

_____

*"God made the world round so we would never
be able to see too far down the road."*
—Isak Dinesen

Once diagnosed with cancer, it becomes easy to scare ourselves by looking down the road and seeing all kinds of horrible situations. Sometimes those painful happenings do occur. But, if we only succeed in paralyzing ourselves with fear and despair, we cannot mobilize our necessary coping abilities and increase the probability that what we fear does not take place. Today, I will focus on enjoying the present rather than worrying about what could happen.

Your thoughts, feelings, and observations today:

Today's Affirmation

The future will take care of itself. I need to enjoy today.

My Personal Affirmation

---

*"The only reason I would take up jogging
is so that I could hear heavy breathing again."*
—Erma Bombeck

Discussion

_____

_____

_____

_____

Your thoughts, feelings, and observations today:

_____

_____

_____

_____

_____

_____

_____

_____

My Personal Affirmation

_____

_____

"I kissed my first girl and smoked my first cigarette on the same day.
I haven't had time for tobacco since."
—Arturo Toscanini

There are lots of choices we can make to affect our health. What is most important about the choices we make is that they help us physically, emotionally, and spiritually feel better. We can literally spend all day worrying about making the right choices and taking care of ourselves. Our choices should be motivated by improved and positive feelings, not fear of dying. I can pick my "vices" consciously. Today, I will focus on positive lifestyle changes I wish to make.

Your thoughts, feelings, and observations today:

_____

_____

_____

_____

_____

_____

_____

_____

Today's Affirmation

I will make lifestyle changes that make me feel better physically, psychologically, and spiritually.

My Personal Affirmation

_____

_____

193

"Yesterday is gone. Tomorrow has not yet come.
We have only today. Let us begin."
—Mother Therese

❋

Discussion

_____

_____

_____

_____

Your thoughts, feelings, and observations today:

_____

_____

_____

_____

_____

_____

_____

_____

_____

My Personal Affirmation
❋

_____

_____

"Cancer and optimism were not considered compatible on this planet."
—Erma Bombeck

✳

The diagnosis of cancer has always carried a multitude of negative associations with it. This seems to put cancer patients in an immediate frame of mind to resign themselves to a horrible fate. Things are changing almost daily and there truly is a new sense of optimism that we are discovering cancer's true secrets, and will develop more sensible ways to treat it. Today, I will take some time to examine some of the new discoveries being made about my cancer to increase my hope about getting well again.

Your thoughts, feelings, and observations today:

_____

_____

_____

_____

_____

_____

_____

_____

Today's Affirmation

I will keep informed of the latest discoveries to conquer cancer so that my attitude remains positive and hopeful.

My Personal Affirmation

✳

_____

_____

"When the past makes you laugh
and you can savor the magic
that let you survive your own war
You'll find that that fire is passion
And there's a door up ahead not a wall."
—Lou Reed, *Magic and Loss* (album)

Discussion

Your thoughts, feelings, and observations today:

My Personal Affirmation

" His entreaties, he says, changed from
'Why me, God?' to 'Show me, God, how to use this experience."
—Gerald Coffee, a Vietnam POW, mentioned in *Remarkable Recovery*
by Caryle Hirshberg & Marc Ian Barasch

While there are a number of aspects of life we can control, there are things that are beyond our control. Our attitude is one aspect of a situation that we can largely control, if we are willing to work at it. Even in the grimmest of situations, people have determined not to let it break their spirit. This is often one of the characteristics of survivors.

Your thoughts, feelings, and observations today:

Today's Affirmation

Today, I will fortify my spirit by examining how I can use this experience of cancer to enrich my life.

My Personal Affirmation

"Did he learn anything when he broke his ankle playing golf?"
"No, not really," I answered.
"So what could he learn from cancer?"
—Rabbi Ben Kamin, *The Path of the Soul*

✳

Discussion

_____

_____

_____

_____

Your thoughts, feelings, and observations today:

_____

_____

_____

_____

_____

_____

_____

_____

_____

My Personal Affirmation

✳

_____

_____

"There is more to life than increasing its speed."
—Gandhi

✠

All too often we get caught up in doing all that we can without allowing time to savor our accomplishments or just enjoy relaxation. Being diagnosed with cancer may make us even more conscious of time. Today, I will make a point to focus on enjoyment rather than productivity.

Your thoughts, feelings, and observations today:

_____

_____

_____

_____

_____

_____

_____

_____

_____

Today's Affirmation

I will slow down to savor more of my life.

My Personal Affirmation

✠

_____

_____

"Sickness will surely take the mind
Where minds don't usually go."
The Who, Tommy (album)
In a dark time, the eye begins to see."
—Theodore Roethke

�֍

Discussion

_____

_____

_____

_____

Your thoughts, feelings, and observations today:

_____

_____

_____

_____

_____

_____

_____

My Personal Affirmation
✷

_____

_____

"In the face of uncertainty, there is nothing wrong with hope."
—O. Carl Simonton

Human beings need hope. When people are hopeless, they quit trying and give up. What we often don't realize is that hope can take many forms. We may all want a remission and that may happen. But, we may have to change the focus of our hope, if that does not happen. Our hope may focus on pain relief, going home from the hospital, finishing business with a loved one, or traveling somewhere we've never been. I will spend some time today examining my sources of hope.

Your thoughts, feelings, and observations today:

_____

_____

_____

_____

_____

_____

_____

_____

Today's Affirmation

My sources of hope will energize me today.

My Personal Affirmation

_____

_____

"If you do not go within, you go without."
—Neale Donald Walsch

✜

Discussion

_____

_____

_____

_____

Your thoughts, feelings, and observations today:

_____

_____

_____

_____

_____

_____

_____

My Personal Affirmation

✜

_____

_____

"Everything's got a moral, if you can find it."
—Lewis Carroll

❈

When someone is diagnosed with cancer, it's common to ask, "Why me?" This is not just a question of anger, but an attempt to make sense of this traumatic event. No one asks for cancer, but once diagnosed, it may be an important message that some things need to change. Today, I will ask myself, "What positive changes have or can this bring?"

Your thoughts, feelings, and observations today:

_____

_____

_____

_____

_____

_____

_____

_____

Today's Affirmation

I will find a positive "moral" to my getting cancer.

My Personal Affirmation

❈

_____

_____

"Who of you by worrying can add a single hour to his (her) life?
Since you cannot do this very little thing, why do you worry about the rest?"
—*Luke* 12:25-26
"Don't worry about the future. Or worry, but know that worrying is as effective
as trying to solve an algebra equation by chewing gum."
—Kurt Vonnegut

✵

Discussion

_____

_____

_____

_____

_____

Your thoughts, feelings, and observations today:

_____

_____

_____

_____

_____

_____

_____

My Personal Affirmation
✵

_____

_____

"The great tragedy of life is not that people perish,
but that they cease to love."
—William Somerset Maugham

✺

Unfortunately, many of us perish before we die. Our hearts harden and we fail to appreciate many of our sources of support and break off other relationships prematurely. Today, I will spend time focusing on different sources of love in my life and new ones I can cultivate.

Your thoughts, feelings, and observations today:

_____

_____

_____

_____

_____

_____

_____

_____

Today's Affirmation

I will open up my heart to old and new sources of love
around me.

My Personal Affirmation

✺

_____

_____

*"Be patient toward all that is unsolved in your heart
...try to love the questions themselves...."*
—Rainer Maria Rilke, *Letters to a Young Poet*

Discussion

_____

_____

_____

_____

_____

Your thoughts, feelings, and observations today:

_____

_____

_____

_____

_____

_____

_____

My Personal Affirmation

_____

_____

"For peace of mind, resign as general manager of the universe."
—Larry Eisenberg

We human beings feel a lot more comfortable when we can know what's ahead. Many times we try to control as much as we can in situations and that is healthy and lowers our anxiety. But there are many times when we try too hard and struggle against situations that are not turning out the way we wanted. Today, I'm going to focus on what I can and can't control about an important situation I am facing.

Your thoughts, feelings, and observations today:

_____

_____

_____

_____

_____

_____

_____

_____

Today's Affirmation

I will control what I can, accept what I can't and learn the difference.

My Personal Affirmation

_____

_____

*"What helps me to go forward is that I stay receptive.
I feel that anything can happen."*
—Anouk Aimee

"Man (woman) never made any material as resilient as the human spirit."
—Bern Williams

✼

Discussion

_____

_____

_____

_____

Your thoughts, feelings, and observations today:

_____

_____

_____

_____

_____

_____

_____

My Personal Affirmation
✼

_____

_____